The early environment in which we grow up has profound, long-lasting, and often irreversible consequences for us throughout our lives. Stresses due to undernutrition in early childhood can mean that in adulthood, individuals are smaller, more prone to disease, and have a shorter life expectancy than those with normal diets. Disease and poor living conditions in infancy and childhood also have profound implications in adulthood. Whilst environmental effects on human growth and development are well documented, the long-term consequences, due to processes taking place at the early stages of growth and development have only recently become the focus of intense study. In this volume, leading researchers in nutrition, epidemiology, human biology, anthropology and physiology bring together a uniquely accessible source of information on this fascinating topic.

SOCIETY FOR THE STUDY OF HUMAN BIOLOGY
SYMPOSIUM SERIES: 37

Long-term Consequences of Early Environment: growth, development
and the lifespan developmental perspective

PUBLISHED SYMPOSIA OF THE
SOCIETY FOR THE STUDY OF HUMAN BIOLOGY

*Numbers 1–9 were published by Pergamon Press, Headington Hill Hall, Headington,
Oxford OX3 0BY. Numbers 10–24 were published by Taylor & Francis Ltd, 10–14 Macklin
Street, London WC2B 5NF. Further details and prices of back-list numbers are available
from the Secretary of the Society for the Study of Human Biology.*

Long-term Consequences of Early Environment: growth, development and the lifespan developmental perspective

37th Symposium Volume of the
Society for the Study of Human Biology

EDITED BY

C.J.K. HENRY

School of Biological and Molecular Sciences
Oxford Brookes University

and

S.J. ULIJASZEK

Department of Biological Anthropology
University of Cambridge

CAMBRIDGE
UNIVERSITY PRESS

CAMBRIDGE UNIVERSITY PRESS
Cambridge, New York, Melbourne, Madrid, Cape Town, Singapore, São Paulo, Delhi

Cambridge University Press
The Edinburgh Building, Cambridge CB2 8RU, UK

Published in the United States of America by Cambridge University Press, New York

www.cambridge.org
Information on this title: www.cambridge.org/9780521118217

First published 1996
This digitally printed version 2009

A catalogue record for this publication is available from the British Library

Library of Congress Cataloguing in Publication data

Long-term consequences of early environment: growth, development, and the
  lifespan developmental perspective/ edited by C.J.K. Henry and S.J. Ulijaszek.
    p.   cm. – (Society for the Study of Human Biology symposium series; 37)
  Includes index.
  ISBN 0 521 47108 7
  1. Human growth – Congresses.   2. Child development – Congresses.
  3. Malnutrition in children – Complications – Congresses.   I. Henry, C.J.K.
  II. Ulijaszek, Stanley J.   III. Series.
  QP84.L646 1996
  612.6 – dc20   95-51685   CIP

ISBN 978-0-521-47108-4 hardback
ISBN 978-0-521-11821-7 paperback

Contents

Dr D.J. Mela
Consumer Sciences Department, Institute of Food Research,
Earley Gate, Reading GR6 6BZ, UK

Dr M.P.M. Richards,
Centre for Family Research, Social and Political Sciences Faculty,
University of Cambridge, Free School Lane, Cambridge CB2 3RF,
UK

Dr L. Rosetta,
Laboratoire de Physiologie des Adaptations, Centre National de la
Recherche Scientifique, Université Paris 5, 24 Rue Faubourg
Saint-Jacques, 75015 Paris, France

Dr S.J. Ulijaszek,
Department of Biological Anthropology, University of Cambridge,
Downing Street, Cambridge CB2 3DZ, UK

Dr S. Wootton
Institute of Human Nutrition, University of Southampton,
Southampton SO16 6YD, UK

Dr C.M. Worthman,
Department of Anthropology, Emory University,
Atlanta, Georgia 30322, USA

Contributors

Professor D.J.P. Barker
Medical Research Council Environmental Epidemiology Unit,
University of Southampton, Southampton General Hospital,
Southampton SO9 4XY, UK

Dr B. Bogin
Department of Behavioral Sciences, University of Michigan at
Dearborn, 4901 Evergreen Road,
Dearborn, Michigan, 48128, USA

Professor N. Cameron
Human Growth Research Programme, Department of Anatomy and
Human Biology, Medical School, University of the Witwatersrand,
Johannesburg, South Africa

Dr S.L. Catt
Consumer Sciences Department, Institute of Food Research,
Earley Gate, Reading RG6 6BZ, UK

Dr R. Garruto
National Institute of Neurological Disorders and Stroke, National
Institutes of Health, Bethesda, Maryland 20892, USA

Professor M.H.N. Golden
Department of Medicine and Therapeutics, University of Aberdeen,
Forrester Hill, Aberdeen AB9 2ZD, UK

Dr A.H. Goodman
School of Natural Science, Hampshire College,
Massachusetts 01002, USA

Dr C.J.K. Henry
School of Biological Sciences, Oxford Brookes University
Gipsy Lane, Headington, Oxford OX3 0BP, UK

Professor A.A. Jackson
Institute of Human Nutrition, University of Southampton,
Southampton SO16 6YD, UK

Acknowledgements

We thank Professor J.C. Waterlow and Professor G.A. Harrison, who shared with the Editors the chairing of the Symposium sessions, and the Kelloggs, Nestlé and Kabi Pharmacia companies for their financial support.

1 Introduction: growth, development and the lifespan developmental perspective

STANLEY J. ULIJASZEK AND C.J.K. HENRY

The Society for the Study of Human Biology Symposium on Long-Term Consequences of Early Environments, held on April 9–10, 1994 at the Pauling Centre, University of Oxford, considered the recent developments in the study of the ways in which early environmental factors can influence human individuals and populations across their lifespan. The topic was examined in a systematic manner by human biologists, auxologists, epidemiologists, anthropologists, physiologists and nutritionists, and the papers discussed at this meeting are published in this volume.

For the past 30 years, human population biology has sought to document and explain processes that have contributed to biological variability in the human species. In particular, it has become clear that the effects of environment upon the genotype are rarely simple, and often not immediately apparent. Environmental effects on human physical and psychological development are considerable and well documented, but the long-term consequences and phenotypic expression in later life of processes taking place during growth and development have only recently become the focus of intense study. Such research has implications for human biology, anthropology, nutrition, clinical science and epidemiology, since many of the long-term outcomes under consideration have consequences for the health and well-being of individuals and populations.

Central to the 'early environment–later outcomes' approach is the understanding that environmental effects on human growth and development, be they nutritional, disease-related, psychosocial or otherwise, can have long-term, often irreversible consequences. Before attempting to address which of the many environmental influences may have irreversible consequences for human form and function, it is important to understand human growth patterns in an evolutionary and life-history context. The

1

authors of the first three chapters, B. Bogin, S. Ulijaszek, and C. Worthman, offer different perspectives of this framework.

The pattern of human growth is characterized by a prolonged period of infant dependency, an extended childhood, and a rapid and large acceleration in growth velocity at adolescence leading to physical and sexual maturation. This trend is of evolutionary significance since it provides the human species with an extended period for brain development, time for the acquisition of technical skills including tool making and food processing, and for socialization and the development of social roles and cultural behaviour. The chapter by B. Bogin examines the implications of this unique growth pattern for biology and behaviour in adolescence and adulthood. The major health problems of the majority of the world's children are either infections or nutritional deficiencies, often leading to severe and costly physical and psychological complications in adulthood. Bogin argues that physically and psychologically well-adjusted adults are more likely to be the outcome of patterns of development which conform to the evolutionarily derived needs of infancy, childhood, and adolescence.

In the following chapter, S.J. Ulijaszek questions the definition of the term 'early environment', and gives an overview of long-term health and disease consequences of some of the relationships between growth and development, nutrition and disease. Ulijaszek suggests that the definition of early environment varies according to the biological phenomenon under investigation, and may be taken to be any time during the course of growth and development. Using examples of non-insulin-dependent diabetes, schizophrenia, muscular development and energetic efficiency, he concludes that the extended period of adult life experienced in industrialized populations relative to past populations and to the human ancestral state allows the greater expression of degenerative diseases. Most long-term outcomes of early environmental influences on growth and development are due either to distorted morphology, or metabolic programming. The former includes phenomena which may arise from impaired development of a somatic structure, while the second consists of phenomena which may be the outcome of physiological 'setting' of some mechanism, probably hormonal, at some critical time in growth and development.

Following from this, C.M. Worthman examines the adaptive processes which shape the human sex ratio. This has implications for the understanding of early environments, since in many societies humans may adjust each successive reproductive event to track existing environmental conditions in ways that can shape the sex ratio and completed fertilities. In high mortality populations, growth curves represent performance of survivors, and males and females often have different mortality risks, both for biological and socio-behavioural reasons. Thus, in many societies, sex and gender differences in early environmental experience may have important implications for child survivorship, the understanding of child growth patterns, demographic structure and the basis of economic production. Factors that drive skewed sex ratios include differential parental costs according to the sex of the child, and the cultural conditions that create sex differences in child mortality. Worthman introduces the idea that the early environment may be different according to sex, may be socially constructed, and may have a temporal dimension.

Many factors influence human growth and development. When considering the growth performance of individuals and populations, healthy outcome is usually assessed by some measure of function, be it physical, physiological, psychological or behavioural. All subsequent chapters consider environmental influences upon some aspect of growth and development which have long-term consequences for some aspect of human function.

In the first of these, N. Cameron reviews literature which shows that birthweight is strongly conditioned by the health and nutritional status of the mother, in that maternal undernutrition, ill-health and other deprivations are the most common causes of retarded fetal growth and/or prematurity. Furthermore, there are intergenerational effects of the antenatal environment on adult physique and disease risk, while the associations between fetal environment and disease risk are well known. Drawing on a variety of sources, including studies of various populations in South Africa, Cameron concludes that birthweight is universally the most important determinant of the newborn surviving to experience healthy growth and development, and of the likelihood of any long-term body size consequence of an adverse antenatal environment.

The notion that catch-up growth, if incomplete, suggests a way whereby smaller adult body size and its functional correlates may be influenced by the environment during infancy and childhood. The view that catch-up growth after an early period of growth faltering can only occur in infancy and childhood has recently been challenged, and M.H.N. Golden presents evidence in opposition to the traditional view. He considers genetic–environmental interactions and their effects on growth and development, particularly in relation to the imprinting of gene expression in offspring by the parental nutritional environment. Golden supports Cameron's position about the importance of the uterine environment for postnatal growth and development, when he states that children stunted postnatally may be able to undergo complete catch-up, while those undergoing growth retardation *in utero* may not.

The idea that diet during early life might influence the rate of growth and development at a later stage is not new. Nor is the concept of the 'plane of nutrition', which involves the regulation of nutritional inputs appropriate to optimal function at various levels including the cellular one. In a largely theoretically oriented chapter, S. Wootton and A. Jackson update and relate these two ideas within the framework of lifespan developmental theory. They develop a model whereby the long-term consequences of nutrition on growth, body size and physiological function are dependent on the programming of metabolic memory, and which includes the process of metabolic change across the lifespan. In support of this model, they describe studies carried out in Southampton which show clear metabolic and dietary behavioural differences between apparently normal but short, and normal and average-statured children. They also cite studies in India which support their view that body size and composition determined in the course of growth and development by interaction with nutritional factors, are likely to play a major role in determining the metabolic characteristics of the body.

There is considerable between- and within-population variability in human nutrient intakes and requirements. Although considerable attention has been paid to defining the range of nutritional requirements at different ages and in different physiological states, almost no attention has been paid to the ways in which the early nutritional environment can shape nutrient needs in later life. In the chapter by C.J.K. Henry, evidence for the existence of such relationships is examined. In particular, the question of whether early manipulation of the nutritional environment of humans can lead to long-term alteration in their nutrient needs later in life due to changes in body composition or organ size, is asked. Using a variety of data from early human studies and animal models, Henry concludes that this may well be the case.

In the chapter which follows, D. Mela and S. Catt examine possible reasons for similarities and differences in food selection. In the context of growth and development, this could be an important factor influencing body size, composition and metabolic characteristics as identified by Wootton and Jackson. Mela and Catt focus on the sensory-affective dimension of food acceptance in relation to the development to taste and smell, and consider evidence for their possible relationships to later food selection. They illustrate the complexity of the dietary environment and of the extent to which sensory acceptance of foods can be associated with the nutritional and physiological properties of foods. Although it is commonly assumed that taste, smell, and food preferences acquired in infancy are maintained through childhood and into adult life, most of the evidence presented in this chapter suggests the opposite. Humans show great plasticity in their food preferences, and it does not appear that specific taste and smell preferences formed early in life track into adulthood.

Perhaps the most dramatic examples of relationships between early environmental impact on growth and development and later biological outcomes are in epidemiology. In the next chapter, D.J.P. Barker reviews recent evidence linking reduced fetal and infant growth with the adult disease. Studies in the UK have shown that babies who were small have, as adults, raised blood pressure, elevated serum cholesterol and plasma fibrinogen concentrations, and impaired glucose tolerance. These are the most important risk factors for coronary heart disease and non-insulin-dependent diabetes mellitus. Barker goes on to describe a model of coronary heart disease causation which puts less emphasis on 'inappropriate lifestyle', and more on environmental factors influencing metabolic programming during growth and development.

It is possible to examine the influence of early environmental factors in past populations, and inferences from such studies may inform our knowledge of more contemporary phenomena. Although models based on the causes and results of physiological disruption or stress have been used to address processual questions in past populations, this approach has only recently been extended to consider the impacts of biological stresses such as undernutrition and infection on child development, and their implications for adult morbidity and mortality patterns. In the chapter by A.H. Goodman, the usefulness of skeletal and other hard tissue memories of early life stress in constructing models of health and disease in past populations is examined.

The development of sexual behaviour differs between males and females, in that the former undergo largely isolated learning of what is desirable in a woman, while the latter initially learn about social relationships, and impose notions of sexuality upon this knowledge. Sex and gender differences in environment, and in social experience across puberty, influence adult sexuality. In a chapter on the childhood environment and the development of sexuality, M.P.M. Richards suggests that existing models assume, on the basis of very little observational data, that sexual behaviour emerges as an inevitable result of the biological changes taking place in puberty. He challenges this notion in the light of recent social research on the development of sexuality. A more realistic view, he claims, is that sexuality has its origins early in life, and continues into adulthood. Puberty, he suggests, should not be viewed as a purely biologically determined event. Rather, it may be seen as a developmental stage whose timing may depend on earlier social relationships, as well as upon nutritional and health circumstances.

Although there is a wealth of information about growth and development in adolescence, and about demographic and biological factors related to adult female fecundity, possible links between the two have not been sought. L. Rosetta reviews the literature on biological and environmental factors influencing the onset and timing of puberty, in relation to the development of fecundity in late adolescence and into adult life. Although it seems likely that there may be some critical period of development during which an impairment of adult reproductive function might be irreversible, to date, none has been identified. Rosetta suggests that research designs involving studies of migrants might be useful in trying to identify such a relationship.

Another area where epidemiological research involving the lifespan developmental perspective might be useful is that of the study of diseases of long latency and slow progression. These include amylotrophic lateral sclerosis, Alzheimer's disease, multiple sclerosis and parkinsonianism. These are thought to be caused by insults expressed through a common pathway leading to similar neuro-degenerative changes over time. How this takes place is unknown, but the type and sequence of events are worthy of research using studies initiated in early life. R.M. Garruto presents current ideas on the progression of such diseases, and concludes that age, timing, and sequence of events are the most important factors influencing the onset of neuro-degenerative disorders.

Collectively, these chapters set as many new questions as they answer. These are mostly concerned with the relationships between growth and development, nutrition, disease, and functional outcome. Other issues include questioning the measurement and definition of environmental factors which can influence growth and development, and the importance of social factors in understanding biological phenomena. The study of human biology from the lifespan developmental perspective as shown in its various forms in this volume, has a future. Whether it will have long-term consequences, and what these might be, remains to be seen.

2 Human growth and development from an evolutionary perspective

BARRY BOGIN

Introduction

The pattern of human growth after birth is characterized by five stages: (i) infancy; (ii) childhood; (iii) juvenile; (iv) adolescence; and (v) adulthood (Bogin, 1988, 1990, 1993). Changes in the velocity of growth from birth to adulthood signal the transitions between these five developmental stages (Figure 2.1). Each of these stages can be defined by distinct biological and behavioural characteristics. Infancy is the period when the mother provides all or some nourishment to her offspring via lactation. Infancy ends when the child is weaned from the breast (or bottle), which in pre-industrialized societies occurs at a median age of 36 months (Detwyller, 1994). Childhood is defined as the stage following weaning, and is a period of time when the youngster is still dependent on older individuals for feeding and protection. Childhood ends when growth of the brain, in weight, is complete. Mathematical modelling of brain growth, using direct measurements from cadavers, indicates that brain growth stops at a mean age of seven years (Cabana, Jolicoeur & Michaud, 1993). The child then progresses to the juvenile stage. Juveniles are defined as, '... prepubertal individuals that are no longer dependent on their mothers (parents) for survival' (Pereira & Altmann, 1985, p. 236). In girls, the juvenile period ends, on average, at about the age of 10 years, two years before it usually ends in boys. The adolescent stage begins with some visible sign of sexual maturation, such as pubic hair, which is followed by development of the other secondary sexual characteristics, a growth spurt in height and weight in both sexes, as well as the onset of adult patterns of sociosexual and economic behaviour. Adolescence ends with the attainment of adult stature which occurs, on average, at about age 18 in women and 21 in men.

In the book *Size and Cycle*, J.T. Bonner (1965) develops the idea that the life cycle of an individual organism, a colony, or a society is, '... the basic unit of natural selection' (p. 52). Bonner discusses organisms as diverse as

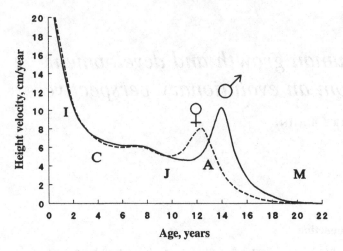

Figure 2.1. Idealized mean velocity curves of growth in height for healthy girls and boys. I-infancy, C-childhood, J-juvenile, A-adolescence, M-mature adult (after Prader, 1984 and other sources).

foraminiferans, slime moulds, algae, volvox, wheat, sequoias, planaria, lichens, blue whales, and colonies of social ants. For each species he shows that its development at different stages of the life cycle, and the duration of each stage, relates to such basic adaptations as locomotion, reproductive rates, and food acquisition. In the 1970s and 1980s many of Bonner's ideas were formalized into the study of life history strategies and evolution. 'A broad definition of life history includes not only the traditional foci such as age-related fecundity and mortality rates, but also the entire sequence of behavioural, physio logical, and morphological changes that an organism passes through during its development from conception to death' (Shea, 1990, p. 325). This chapter focuses on human childhood, in terms of both its evolution and its place in human life history strategy.

The evolution of childhood

Figure 2.2 represents a summary of the evolution of the human pattern of growth and development (the evolution of adolescence is not discussed in this chapter, but see Bogin, 1993, 1994*a,b*). This figure must be considered as 'a work in progress', as only the data for the first and last species (*Pan* and *Homo sapiens*) are known with some certainty. The patterns of growth of the fossil hominid species are reconstructions based on the traditional methods of human palaeontology: comparative anatomy, comparative physiology, comparative ethology, archaeology, as well as some specula-tion. In particular, the work of R.D. Martin and colleagues (1983; Harvey,

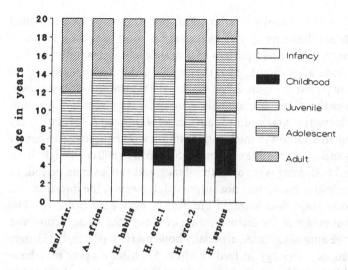

Figure 2.2. The evolution of hominid life history during the first 20 years of life. Abbreviated nomenclature as follows: A.afar.-*Australopithecus afarensis*, A. africa – *Australopithecus africanus*, H. habilis – *Homo habilis*, H. erec. 1 – early *Homo erectus*, H. erec. 2 – late *Homo erectus*, H. sapiens – *Homo sapiens*.

Martin & Clutton-Brock, 1987) on comparative patterns of brain growth in apes and humans is utilized. Martin shows that apes have a pattern of brain growth that is rapid before birth and relatively slower after birth. In contrast, humans have rapid growth both before and after birth. Martin also notes that, from newborn to adult, apes slightly more than double their brain size – a multiplier of 2.3; whereas humans more than triple their brain size – a multiplier of 3.5. Finally, Martin argues that an adult brain size of 873 cm³ or larger represents a 'cerebral rubicon', separating ape-like and human-like patterns of brain growth. The 873 cm³ value is calculated by multiplying the average human infant brain size of 384 cm³ by the ape multiplier of 2.3.

Martin's analysis is elegant and tenable; nevertheless, the difference between ape and human brain growth is not only a matter of velocity; it is also a matter of life history stages. Brain growth for both apes and human beings ends at the start of the juvenile stage, which means that apes complete their brain growth during infancy. Human beings, however, insert the childhood stage between the infant and juvenile stages. Childhood may provide the time necessary to grow the larger human brain. Following this line of reasoning, any fossil human, or any of our fossil hominid ancestors, with an adult brain size above Martin's 'cerebral rubicon' may have included a childhood stage of growth as part of its life history.

Australopithecus afarensis is a hominid, but shares many anatomical features with non-hominid species including an adult brain size of about 400 cm³ (Simons, 1989) and a pattern of dental development indistinguishable from extant apes (Smith, 1991). Therefore, the chimpanzee and *A. afarensis* are depicted in Figure 2.2 as sharing the typical tripartite stages of postnatal growth of social mammals – infant, juvenile, adult (Pereira, 1993). To achieve the larger adult brain size of *A. africanus* (442 cm³) may have required an addition to the length of infancy. The rapid expansion of adult brain size during the time of *Homo habilis* (650 to 800 cm³) might have been achieved with expansion of both infancy and the juvenile period, as Martin's 'cerebral rubicon' was not surpassed. However, the insertion of a brief childhood stage into hominid life history may have occurred. The archaeological evidence for intensification of stone tool manufacture and use to scavenge animal carcasses, especially bone marrow (Potts, 1988), may be interpreted as a strategy to feed children. Such scavenging may have been needed to provide the protein, some of the minerals, and the fat (a dense source of energy) that children require for growth of the brain and body (Leonard & Robertson, 1992).

Further brain size increase occurred during *H. erectus* times. The earliest adult specimens have brain size of 850 to 900 cm³. This places *H. erectus* at or above Martin's 'cerebral rubicon' and may justify an expansion of the childhood period to provide the high quality foods needed for the rapid, human-like, pattern of brain growth. It should be noted from Figure 2.2 that, from *Australopithecus* to *H. erectus*, the infancy period shrinks as the childhood stage expands. As will be discussed below, this gave *H. erectus*, and all later hominids, a reproductive advantage over all other hominoids. Later *H. erectus*, with adult brain sizes up to 1100 cm³, are depicted with further expansion of childhood and the insertion of the adolescent stage. In addition to bigger brains, later *H. erectus* shows increased complexity of technology (tools, fire, and shelter) and social organization that were likely correlates of the biology and behavior associated with further development of the childhood stage. The transition to archaic and finally modern *H. sapiens* expands the childhood stage to its current dimension.

Why do human beings have childhood?

Brain size of extant and fossil hominoids provides some idea of when human life stages may have evolved, but does not explain why they evolved. To make sense of the pattern of human growth, one must look for the 'basic adaptations' that Bonner describes. The most basic of these adaptations are those that relate to evolutionary success. This is traditionally measured in terms of the number of offspring that survive and reproduce. Biological and behavioural traits do not evolve unless they confer upon their owners some degree of reproductive advantage, in terms of survivors a generation or more later. Bogin (1988, pp. 74–75) lists seven reasons for the evolution of human childhood from the perspective of reproductive success. The first three are the traditional 'textbook' explanations that emphasize learning, an idea that goes back to Spencer (1886). These are:

1. an extended period for brain growth,
2. time for the acquisition of technical skills, e.g. toolmaking and food processing, and
3. time for socialization, play, and the development of complex social roles and cultural behaviour.

These reasons are valid in as much as they confer an advantage to pre-adult individuals. However, this brain-learning list of explanations cannot account for the initial impetus for the insertion of childhood into human life history. A childhood stage of development is not necessary for the type of learning listed here. The prolonged infancy and juvenile period of the social carnivores (Bekoff & Beyers, 1985) and apes (Bogin, 1994a) can serve that function. Rather, childhood may be better viewed as a reproductive adaptation for the parents of the child, as a strategy to minimize the risks of starvation and predation, and as a mechanism that allows for more precise 'tracking' of ecological conditions via developmental plasticity. These reproductive, survival, and plasticity advantages of childhood are discussed, respectively, in the following sections.

Human reproductive strategy

There are limits to the amount of delay between birth and sexual maturity that any species can tolerate. The Great Apes are examples of this limit. Chimpanzee females in the wild reach menarche (the first menstruation) at 11 to 12 years of age and have their first births at an average age of 14 years (Goodall, 1983). The average period between successful births in the wild is 5.6 years, and young chimpanzees are dependent on their mothers for about five years (Teleki, Hunt & Pfifferling, 1976; Goodall, 1983). Even though it is estimated that about 35% of all live-born chimpanzees survive to their mid-twenties, a significantly greater percentage of survival than most other species of animals, the chimpanzee is at a reproductive threshold. Goodall, for example, reports that, for the period 1965 to 1980, there were 51 births and 49 deaths in one community of wild chimpanzees. Galdikas and Wood (1990) present data for the orangutan which shows that these apes are in a more precarious situation. Compared with the 5.6 years between successful births of chimpanzees, the orangutan female waits

7.7 years. Lovejoy (1981) argues that chimpanzees and other apes are just barely able to maintain population size, a situation he calls a 'demographic dilemma' (p. 211).

The great apes, and early hominids such as *A. africanus* or *H. habilis*, reached this demographic dilemma by extending infancy, forcing a demand on nursing to its limit. Human beings overcome this predicament by reducing the length of infancy and inserting childhood between the end of infancy and the juvenile period. Free from the demands of nursing and the physiological brake that nursing places on ovulation (Ellison, 1990), the human mother can reproduce soon after her infant becomes a child. An oft cited example, the !Kung, are a traditional hunting and gathering society of southern Africa. A !Kung woman's age at her first birth averages 19 years and subsequent births follow about every 3.6 years, resulting in an average fertility rate of 4.7 children per woman (Howell 1976, 1979, Short, 1976). Women in another hunter–gatherer society, the Hadza (Blurton-Jones *et al.*, 1992), have even shorter intervals between successful births, stop nursing about one year earlier, and average 6.15 births per woman.

These relationships are illustrated in Figure 2.3, a comparison of several life history events for female great apes and human beings. The data are drawn from studies of wild-living animals for the great apes and non-contracepting hunting and gathering or horticultural populations for human beings. The infancy dependency period for each of the apes species is longer than that for humans. Ape infancy ends after eruption of the first permanent molar, which is probably a requirement so that the juvenile ape can acquire and process foods of the adult diet. Human infancy ends before eruption of the first permanent molar, that is, before the youngster can process adult foods. The evolution of childhood as a stage in human life history 'fills the gap', between the infant's dependency on the mother for food via nursing and the feeding independence of the juvenile. The addition of a childhood stage and the prolongation of the juvenile and adolescent stages of development of humans also delays the age of both menarche and first birth. However, compared with the great apes, humans reduce the birth spacing interval and therefore, each women may produce more offspring during her life than any female ape. This results in an increase in reproductive fitness if the additional offspring survive to maturity. !Kung, Hadza, and all human parents help to ensure survival of their offspring by providing all their children with food, not just their current infant, for a decade or longer. The child must be given foods that are specially chosen and prepared and these may be provided by older juveniles, adolescents, or adults. In Hadza society, for example, grand-mothers and great-aunts are observed to supply a significant amount of food to children (Blurton-Jones, 1993). Lancaster and Lancaster (1983)

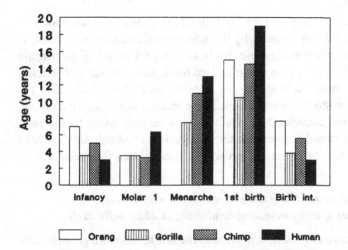

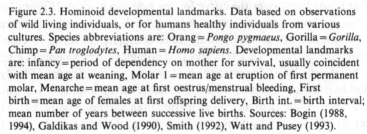

Figure 2.3. Hominoid developmental landmarks. Data based on observations
of wild living individuals, or for humans healthy individuals from various
cultures. Species abbreviations are: Orang = *Pongo pygmaeus*, Gorilla = *Gorilla*,
Chimp = *Pan troglodytes*, Human = *Homo sapiens*. Developmental landmarks
are: infancy = period of dependency on mother for survival, usually coincident
with mean age at weaning, Molar 1 = mean age at eruption of first permanent
molar, Menarche = mean age at first oestrus/menstrual bleeding, First
birth = mean age of females at first offspring delivery, Birth int. = birth interval;
mean number of years between successive live births. Sources: Bogin (1988,
1994), Galdikas and Wood (1990), Smith (1992), Watt and Pusey (1993).

call this type of offspring investment 'the hominid adaptation', for no
other primate or mammal does this.

Given this unique parental behaviour, it may be more profitable to view
the advantages of childhood from the perspective of adults. This leads to
the following four additional reasons for the evolution of childhood (these
replace points 4 to 7 in Bogin, 1988, p. 75):

4. A childhood growth stage may have originally evolved as a feeding adaptation

Providing children with food frees the mother from the demands of nursing
and the resultant inhibition of ovulation related to continuous nursing.
This decreases the interbirth interval and increases reproductive fitness.
Such provisioning, however, is effective only if older individuals provide
foods appropriate to the needs of the childhood and, as well, feel compelled
to meet these needs. Appropriate weaning foods are required because the
small gastrointestinal tracts of young children are not capable of digesting
sufficient quantities of the adult diet to meet all nutritional requirements,

especially energy (Bogin, 1988, p. 130). This problem is alleviated in contemporary human societies by the selection of calorie dense foods, as well as by using technology (e.g. tools and fire for cooking) to prepare weaning foods – behaviours first practised by *H. habilis* or early *H. erectus*. Alternatively, older individuals can provide adult-type foods that are partially predigested by mastication. These feeding strategies for children, however, require considerable investment of time and energy by the older individuals. A stimulus to release these behaviours towards children may be found in the very pattern of growth of the children themselves.

5. The allometry of the growth of the human child releases nurturing and care-giving behaviours in older individuals

The central nervous system, in particular the brain, follows a growth curve that is advanced over the curve for the body as a whole (Figure 2.4). The brain achieves adult size when body growth is only 40% complete, and reproductive maturation is only 10% complete. A series of ethological observations (Lorenz, 1971) and psychological experiments (Todd *et al.*, 1980; Alley, 1983) demonstrate that these growth patterns of body, face, and brain allow the human child to maintain a neotenous (i.e. 'cute') appearance longer than any other mammalian species (Figure 2.5). The perception of neoteny by adults facilities parental investment by maintaining the potential for nurturing behaviour of adults towards infants and older (but still physically dependent) children (Bogin, 1988, pp. 98–104; 1990).

6. The relatively slow rate of body growth and small body size of children reduces competition with adults for food resources

Slow-growing, small children require less total food than bigger juveniles, adolescents, and adults. A child on the fifth centile of the NCHS reference curves for growth, for example, requires 17% less dietary energy for maintenance than a child on the 50th centile (Ulijaszek & Strickland, 1993). Thus, provisioning children, though time consuming, is not as onerous a task of investment as it would be, for instance, if both brain and body growth were rapid simultaneously.

Similarly, the task of child care becomes even less onerous because:

7. Children do not require nursing

Early neurological maturity versus late sexual maturity allows juveniles and young adolescents to provide much of their own care and also provide care for children (Bogin, 1994a). Again, this frees adults, especially the

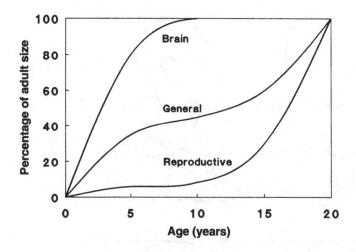

Figure 2.4. Growth curves for different body tissues. The 'general' curve represents growth in stature or total body weight. The 'brain' curve is for total brain weight and the 'reproductive' curve represents the weight of the gonads and primary reproductive organs (after Scammon, 1930 with brain growth data from Cabana *et al.*, 1993).

mother, for subsistence activity, adult social behaviours, and further childbearing.

A further important reason for the evolution of childhood is that:

8. Childhood allows for developmental plasticity

Following the discussion in Stearns (1992, p. 62), the term plasticity means a change in the phenotype of the individual caused by a change in the environment. The fitness of a given phenotype varies across the range of variation of an environment. When phenotypes are fixed early in development, such as in mammals that mature sexually soon after weaning (e.g. rodents), environmental change and high mortality are positively correlated. Social mammals (carnivores, elephants, primates) prolong the developmental period by adding a juvenile stage between infancy and adulthood. Adult phenotypes develop more slowly in these mammals. They experience a wider range of environmental variation, and the result is a better conformation between the individual and the environment. Fitness is increased in that more offspring survive to reproductive age than in mammalian species without a juvenile stage (Lancaster & Lancaster, 1983). Human beings insert the childhood stage between infancy and the juvenile

Figure 2.5. The releasing schema for human parental care responses. Left: head proportions perceived as 'lovable' (child, jeroba, Pekinese dog, robin). Right: related heads which do not elicit parental drive (man, hare, hound, golden oriole). Note that, except for the child, all heads depict adult animals (from Lorenz, 1971).

period. This results in an additional four years of relatively slow physical growth and allows for behavioural experience that further enhances developmental plasticity. The combined result is increased fitness. By comparison, humans in traditional societies rear about 50% of their offspring to adulthood, whereas monkeys and apes rear between 12 and 36% of offspring to adulthood.

Risks of childhood

The preceding list of the advantages of childhood needs to be tempered against the hazards of this developmental stage. Dependency on older individuals for food and protection, small body size, slow rate of growth, and delayed reproductive maturation each entail liabilities to the child. Point (6) in the list of advantages, for example, alludes to the risk from competition with other members of the social group for food and other resources. In this case, and in others, the neotenous 'charms' of childhood do not provide for total security.

To illustrate this point, one can examine traditional societies of both historic and prehistoric eras. In such societies, including hunter–gatherers and horticulturalists, about 35% of live born humans die by age 7, that is, by the end of childhood. Even if two-thirds of these deaths occur during the infancy stage, the childhood period still has an appreciable risk for death. In hunter–gatherer societies, starvation, accidents, and predation account for most childhood deaths. !Kung parents, in an effort to minimize these risks, protect their children by confining them to camp and the watchfulness of adults (Draper, 1976). Mbuti parents (hunter–gatherers of central African forests) allow children and juveniles to form age-graded play groups, also located near the camp. This method of grouping children with older juveniles and adolescents demands less parental supervision (Turnbull, 1983). However, the risks of hunter–gatherer childhood can, perhaps, best be illustrated by Hadza society. Hadza children aged three to eight years form age-graded work/play groups that may wander far from camp, often out of sight of older individuals, and carry out costly and dangerous tasks including significant foraging for food (Blurton-Jones, 1993). One study finds that, of 301 Hadza infants born to 75 women aged 20 to 50+ years old, 115 (39%) of the offspring had died by the time of the survey (Blurton-Jones *et al.*, 1992). Demographic modelling indicates that most of the deaths occurred during infancy and childhood.

Today, hunter–gatherer and traditional horticultural societies account for less than 1% of human cultures, so it may be more instructive to examine current risks to children in post-Colonial and industrialized societies. Mild-to-moderate energy undernutrition is, perhaps, the most common risk. Worldwide, estimates are that between 60 to 75% of all children are undernourished (UNICEF, 1985). Malnutrition may be due to food shortages alone, but equally likely they are due to work loads and infectious disease loads placed on children that compromise their energy balance (Worthman, 1993).

Another risk for children in the contemporary world is that of abuse and neglect. One estimate of the worldwide mortality from abuse and neglect is between 13 and 20 infants and children per 1000 live births (Belsey, 1993). The incidence of all suffering from abuse and neglect is probably higher. One reason for child abuse and neglect is that the biology of childhood may not keep pace with the rapidity of technological, social, and ideological change relating to families and their children. It is now technologically possible to nourish infants without breast feeding and this allows parents (mothers) an opportunity to pursue other economic activities, or to have another baby. Among the poor populations of the developing countries, short birth intervals (less than 23 months) compromise the health of both the infant and the mother (Huttly *et al.*, 1992). A major negative effect on the infant is low birthweight which is known to impair both physical growth and cognitive development during childhood and later life stages (Garn, Pesick & Pilkington, 1984). In populations of the more developed nations, such as among the United States middle-class, fewer than 20% of infants are breast fed and the weaning process – from bottles and formula – may begin by three months of age (Detwyller, 1994). This severely curtails the infancy stage and prolongs childhood. These 'premature children' present a problem for care, as they are still biologically within the infancy stage of development. The problem is often solved by sequestering these young to restraining devices such as high chairs, playpens, and cribs or segregating them from the family by placement in creches or preschools. When the infants react poorly to these arrangements, the frustrated parents or care givers may respond with abusive or neglectful behaviour.

There are long-term consequences to the hunger, work, disease, and abuse or neglect that children suffer. There are physical growth effects, including short stature, inadequate development of muscle and body fat tissue, altered body proportions (especially relatively short legs), and delayed sexual maturation. These sequelae are linked with reduced work capacity and earning power later in life, which, in turn, are linked to further hunger and disease risk in the next generation of children (Bogin, 1988, pp. 126–159). The literature is replete with examples of the deleterious psychological and emotional effects of physical, sexual, and emotional abuse during childhood (Moeller, Bachmann & Moeller, 1993). There are also lasting physical consequences as found in one study of 668 middle-class American women. About half the sample reported some type of abuse as children, and these same women had significantly more hospitalizations for illness and lower ratings of their overall health than the women not reporting childhood abuse (Moeller *et al.*, 1993). Moreover, the women who had been abused as children were more likely also to experience abuse as adults. Thus, childhood abuse and neglect have lasting consequences for adult resiliency against psychological and physical pathology.

A provocative hypothesis linking childhood abuse and neglect with reproductive biology was proposed by Belsky and colleagues (1991). In an attempt to develop an evolutionary theory of early socialization and later reproductive strategy, they describe two alternative developmental pathways. The first alternative, '. . . is characterized, in childhood, by a stressful rearing environment and the development of insecure attachments to parents and subsequent behavior problems; in adolescence by early pubertal development and precocious sexuality; and, in adulthood by unstable pair bonds and limited investment in child rearing' (p. 647). The second pathway '. . . is characterized by the opposite' (p. 647). Belsky and colleagues (Moffitt *et al.*, 1992) tested their hypothesis using data from a longitudinal study that followed a sample of girls in New Zealand until age 16 years. Family conflict and absence of the father from the family occurring during a girl's childhood are associated with an earlier age of menarche – the measure of sexual development used in this study. However, age of menarche is more parsimoniously predicted by a genetic inheritance model (i.e. mother's age at menarche) than by any of the behavioural data. A similar finding is reported by Campbell and Udry (1995). Based on a retrospective analysis of 1001 girls participating in the California Childhood Health and Development Study, only the mother's age at menarche and mother's level of education were significant predictors of the daughter's age of menarche. No measure of childhood stress is a significant predictor of menarche. Follow-up is required to see how the childhood experience of these girls relates to their adult relationships and parental investment.

The political economy of childhood

The benefits, risks, and the long-term consequences of childhood are played out in social and economic environments. Most of the hunger, disease, and heavy work loads that children suffer are linked with poverty. Much anthropological, historical, and economic research on children takes the socioeconomic environment into account (Bielicki, 1986; Bogin, 1988, 1995; Fogel, 1986). Few anthropological studies on children, however, explicitly set these socioeconomic environments in their political context. This omission is being criticized, and addressed, by Thomas and colleagues who specifically examine the 'biology of poverty' (Goodman *et al.*, 1988; Leatherman, 1992; Thomas, 1992). One of their more important findings for the field of human biology in general is that the emphasis on high altitude hypoxia as the primary cause of poor child growth in the Andean *altiplano* has been wrongly placed. Rather, political inequalities between ethnic groups lead to differences in access to essential resources such as land, education, wage labour result in poverty and poor nutrition, and heavy work loads for children in most high altitude populations (Leonard *et al.*, 1990). This conclusion is supported by other research groups (de Meer, Bergman & Kusner, 1993).

A few additional examples of the effects of political economy and poverty on children are briefly reviewed here. Schell (1992) develops a risk model for lead exposure among poor, African–American children in United States cities. The model focuses on the political and economic forces that segregate some groups of children into 'urban lead belts', and also focuses on the negative cognitive and behavioural effects of lead exposure, such as poor school performance.

Children are especially vulnerable to the effects of warfare owing to their dependency as well as their physical and psychological sensitivity (plasticity) to environmental disturbances. Civil war in Guatemala between 1979 and 1982 displaced more than 250 000 people from their homes, many of whom migrated to Mexico and the United States. Bogin (1995) examines Maya children and youth, between 4 and 12 years old ($n = 240$), now living in California and Florida. Living under relatively greater economic and political freedom in the United States, the migrant children are significantly taller, heavier, and carry more fat and muscle mass than Mayan children living in a traditional village in Guatemala.

How the Mayan immigrants will fare in the future is far from certain, for political policies of the past decade have compromised the United States national environment for child development. Following declines in the poverty rate for children in the 1960s and 1970s, that rate increased during the 1980s (Segal, 1991). 'In addition, although more children became poor, social service provision did not increase and in many instances was cut back' (Segal, 1991, p. 454). An increase in homelessness for children and their families is one result of these policies. One survey of 146 such families residing in homeless shelters assessed the health of the children (Parker *et al.*, 1991). Compared with national data, the homeless children were found to have high rates of accidents and injuries, especially burns, and lead toxicity. They also evidenced delays in receptive vocabulary and visual motor skills. The long-term consequences of these findings are not certain.

Conclusions

Childhood is a unique stage in the life history of human beings. There is some evidence that childhood evolved nearly two million years ago, perhaps during the time of *Homo habilis*. It is certainly true that childhood provides 'extra' time for brain development and learning. However, its initial selective value may be more closely related to parental strategies to give birth to new offspring and provide care for existing dependent young. Moreover, the childhood stage allows for increased developmental plasticity and fitness for the young. The dependency of childhood also entails risks. Deviant behaviour by older individuals, social groups, and political institutions can adversely and permanently affect children.

A troubled childhood has long-term consequences. The problems of adults who were abused or neglected as children are often cited. Less openly discussed are undernutrition in the poor nations and overnutrition in the wealthy nations. These are, arguably, the major health threats to the world's children. Both types of malnutrition often lead to severe and costly physical and psychological complications in adulthood. To paraphrase William Wordsworth, the child is father and mother to the adult. It is certainly easier to produce physically healthy and psychologically well-adjusted adults if their development conforms to the evolutionarily derived needs of infancy, childhood, and the other stages of human life history.

Acknowledgements

The author thanks Mary Ann Wnetrzak and Stanley Ulijaszek for reading drafts of this chapter and offering many helpful criticisms. Financial support to attend the SSHB meetings in Oxford, UK was kindly provided by the Society for the Study of Human Biology, Dr C.G.N. Mascie-Taylor, and Pharmacia Peptide Hormones, Stockholm.

References

Alley, T.R. (1983). Growth-produced changes in body shape and size as determinants of perceived age and adult caregiving. *Child Development*, **54**, 241–8.

Bekoff, M. & Beyers, J.A. (1985). The development of behavior from evolutionary and ecological perspectives in mammals and birds. *Evolutionary Biology*, **19**, 215–86.

Belsey, M.A. (1993). Child abuse: measuring a global problem. *World Health Stat. Q.* **46**, 69–77.

Belsky, J., Steinberg, L. & Draper, P. (1991). Childhood experience, interpersonal development, and reproductive strategy: an evolutionary theory of socialization. *Child Development*, **62**, 647–70.

Bielicki, T. (1986). Physical growth as a measure of the economic wellbeing of populations: the twentieth century. In *Human Growth*, ed. F. Falkner, J.M. Tanner, 2nd ed., Vol. 3, pp. 283–305. New York: Plenum.

Blurton-Jones, N.G. (1993). The lives of hunter–gatherer children: effects of parental behavior and parental reproductive strategy. In *Juvenile Primates*, ed. M.E. Pereira and L.A. Fairbanks, pp. 309–26. Oxford: Oxford University Press.

Blurton-Jones, N.G., Smith, L.C., O'Connell, J.F. & Handler, J.S. (1992). Demography of the Hadza, an increasing and high density population of savanna foragers. *American Journal of Physical Anthropology*, **89**, 159–81.

Bogin, B. (1988). *Patterns of Human Growth*. New York: Cambridge University Press.

Bogin, B. (1990). The evolution of human childhood. *BioScience*, **40**, 16–25.

Bogin, B. (1993). Why must I be a teenager at all? *New Scientist*, **137**(Mar. 6), 34–8.

Bogin, B. (1994a). The evolution of learning. In *The International Encyclopedia of Education*, ed. T. Husen and T.N. Postlethwaite, 2nd ed., Oxford: Pergamon Press, (in press).

Bogin, B. (1994b). Adolescence in evolutionary perspective. *Acta Paediatrica*, Suppl. **406**, 29–35.

Bogin, B. (1995). Plasticity in the growth of Mayan refugee children living in the United States. In *Human Variability and Plasticity*. ed. C.G.N. Mascie-Taylor and B. Bogin, pp. 46–74. Cambridge: Cambridge University Press.

Bonner, J.T. (1965). *Size and Cycle*. Princeton, New Jersey: Princeton University Press.

Cabana, T., Jolicoeur, P. & Michaud, J. (1993). Prenatal and postnatal growth and allometry of stature, head circumference, and brain weight in Québec children. *American Journal of Human Biology*, **5**; 93–9.

Campbell, B. & Udry, J.R. (1995). Mother's age at menarche, not stress, accounts for daughter's age at menarche. *Journal of Biosocial Science* **27**, 359–68.

de Meer, K., Bergman, R. & Kusner, J.S. (1993). Differences in physical growth of Aymara and Quechua children living at high altitude in Peru. *American Journal of Physical Anthropology*, **90**, 59–75.

Detwyller, K.A. (1994). A time to wean: the hominid blueprint for the natural age of weaning in modern human populations (abstract). *American Journal of Physical Anthropology*, Suppl. **18**, 80.

Draper, P. (1976). Social and economic constraints on child life among the !Kung. In *Kalahari Hunter Gatherers*, ed. R.B. Lee, E. DeVore, pp. 199–217. Cambridge, Mass.: Harvard University Press.

Ellison, P.T. (1990). Human ovarian function and reproductive ecology: new hypotheses. *American Anthropology*, **92**, 933–52.

Fogel, R.W. (1986). Physical growth as a measure of the economic wellbeing of populations: the eighteenth and nineteenth centuries. In *Human Growth*, ed. F. Falkner and J.M. Tanner, 2nd ed., Vol. 3, pp. 263–81. New York: Plenum.

Galdikas, B.M., Wood, J.W. (1990). Birth spacing patterns in humans and apes. *American Journal of Physical Anthropology*, **83**, 185–91.

Garn, S.M. Pesick, S.D. & Pilkington, J.J. (1984). The interaction between prenatal and socioeconomic effects on growth and development on childhood. In *Human Growth and Development*, ed. Borms, J., Hauspie, R., Sand, A., Susanne, C. and Hebbelinck, M., pp. 59–70. New York: Plenum.

Goodall, J. (1983). Population dynamics during a 15 year period in one community of free-living chimpanzees in the Gombe National Park, Tanzania. *Zeitschrift Tierpsychologie*, **61**, 1–60.

Goodman, A.H., Thomas, R.B., Swedlund, A.C. & Armelagos, G.A. (1988). Biocultural perspectives on stress in prehistoric, historical, and contemporary population research. *Yearbook of Physical Anthropology*, **31**, 169–202.

Harvey, P.H., Martin, R.D. & Clutton-Brock, T.H. (1987). Life histories in comparative perspective. In *Primate Societies* ed. B.B. Smuts, D.L. Cheney, R.M. Seyfarth, R.W. Wrangham & T.T. Struhsaker, pp. 181–96. Chicago: University of Chicago Press.

Howell, N. (1976). The population of the Dobe area !Kung. In *Kalahari Hunters-Gatherers*, ed. R.B. Lee, I. DeVore, pp. 137–57. Cambridge: Harvard University Press.

Howell, N. (1979). *Demography of the Dobe !Kung*. New York, Academic Press.

Huttley, S.R., Victoria, C.G., Barros, F.C. & Vaughn, J.P. (1992). Birth spacing and child health in urban Brazilian children. *Pediatrics* **89**, 1049–54.

Lancaster, J.B. & Lancaster, C.S. (1983). Parental investment: The hominid adaptation. In *How Humans Adapt*. ed. D.J. Ortner, pp. 33–65. Washington D.C.: Smithsonian Institution Press.

Leatherman, T.L. (1992). Adaptation as a dialectical process: health and household economy in southern Peru. Paper presented at 'Political-Economic Perspectives in Biological Anthropology,' Wenner-Gren Foundation for Anthropological Research symposium, Cabo San Lucas, Mexico.

Leonard, W.R., Leatherman, T.L., Carey, J.W. & Thomas, R.B. (1990). Contributions of nutrition versus hypoxia to growth in rural Andean Populations. *American Journal Human Biology*, **2**, 613–26.

Leonard, W.R., Robertson, M.L. (1992). Nutritional requirements and human evolution: a bioenergetics model. *American Journal of Human Biology*, **4**, 179–95.

Lorenz, K. (1971). Part and parcel in animal and human societies: a methodological discussion. In *Studies in Animal and Human Behavior*. (ed. and trans.) R. Martin, Vol. 2, pp. 115–95. Cambridge: Harvard University Press.

Lovejoy, C.O. (1981). The origin of man. *Science*, **211**, 341–50.

Martin, R.D. (1983). Human brain evolution in an ecological context. Fifty-second James Arthur Lecture, American Museum of Natural History, New York.

Moeller, T.P., Bachmann, G.A. & Moeller, J.R. (1993). The combined effects of physical, sexual, and emotional abuse during childhood: long-term health consequences for women. *Child Abuse Negl.*, **17**, 623–40.

Moffitt, T.E., Caspi, A., Belsky, J. & Silva, P.A. (1992). Childhood experience and the onset of menarche: a test of a sociobiological model. *Child Development*, **63**, 47–58.

Parker, R.M., Rescorla, L.A., Finkelstein, J.A., Barnes, N., Holmes, J.H. & Stolley, P.D. (1991). A survey of the health of homeless children in Philadelphia shelters. *American Journal of Diseases in Childhood*, **145**, 520–6.

Pereira, M.E. (1993). Evolution of the juvenile period in mammals. In *Juvenile Primates*, ed. M.E. Pereira and L.A. Fairbanks, pp. 17–27. Oxford: Oxford University Press.

Pereira, M.E., Altmann, J. (1985). Development of social behavior in free-living nonhuman primates. In *Nonhuman Primate Models for Human Growth and Development*, ed. E.S. Watts, pp. 217–309. New York: Alan R. Liss.

Potts, R. (1988). *Early Hominid Activities at Olduvai*. New York: Aldine de Gruyter.

Prader, A. (1984). Biomedical and endocrinological aspects of normal growth and development. In *Human Growth and Development*, ed. J. Borms, R. Hauspie, A. Sand, C. Susanne, M. Hebbelinck, pp. 1–22. New York: Plenum.

Scammon, R.E. (1930). The measurement of the body in childhood. In *The Measurement of Man*, ed. J.A. Harris *et al.*, pp. 173–215. Minneapolis: University of Minnesota Press.

Schell, L.M. (1992). Risk focusing: an example of biocultural interaction. In R. Huss-Ashmore, J. Schall and M. Hediger (eds) *Health and Lifestyle Change. MASCA Research Papers in Science and Archaeology*, **9**, pp. 137–44.

Segal, E.A. (1991). The juvenilization of poverty in the 1980s. *Soc. Work*, **36**, 454–7.

Shea, B.T. (1990). Dynamic morphology: growth, life history, and ecology in primate evolution. In *Primate Life History and Evolution*, ed. C.J. DeRousseau, pp. 325–52. New York: Wiley-Liss.

Short, R.V. (1976). The evolution of human reproduction. *Proceedings of the Royal Society, Series B*, **195**, 3–24.

Simons, E.L. (1989). Human origins. *Science*, **245**, 1343–50.

Smith, B.H. (1991). Dental development and the evolution of life history in Hominidae. *American Journal of Physical Anthropology*, **86**, 157–74.

Smith, B.H. (1992). Life history and the evolution of human maturation. *Evolutionary Anthropology*, **1**, 134–42.

Spencer, H. (1886). *The Principles of Biology*, Vols. I and II. New York: D. Appleton.

Stearns, S.C. (1992). *The Evolution of Life Histories*. Oxford: Oxford University Press.

Teleki, G.E., Hunt, E., Pfifferling, J.H. (1976). Demographic observations (1963–1973) on the chimpanzees of the Gombe National Park, Tanzania. *Journal of Human Evolution*, **5**, 559–98.

Thomas, R.B. (1992). The biology of poverty. Paper presented at 'Political–Economic Perspectives in Biological Anthropology,' Wenner-Gren Foundation for Anthropological Research symposium, Cabo San Lucas, Mexico.

Todd, J.T., Mark, L.S., Shaw, R.E., Pittenger, J.B. (1980). The perception of human growth. *Scientific American*, **242**, 132–44.

Turnbull, C.M. (1983). *The Mbuti Pygmies*. New York: Holt, Rinehart, Winston.

Ulijaszek, S.J., Strickland, S.S. (1993). *Nutritional Anthropology: Prospects and Perspectives*. London: Smith Gordon.

UNICEF (1985). *The State of the World's Children*. New York: Oxford University Press.

Watts, D.P., Pusey, A.E. (1993). Behavior of juvenile and adolescent great apes. In *Juvenile Primates*, ed. M.E. Pereira and L.A. Fairbanks, pp. 148–70. Oxford: Oxford University Press.

Worthman, C. (1993). Biocultural interactions in human development. In *Juvenile Primates*, ed. M.E. Pereira and L.A. Fairbanks, pp. 339–58. Oxford: Oxford University Press.

3 Long-term consequences of early environmental influences on human growth: a developmental perspective

STANLEY J. ULIJASZEK

Introduction

For more than 30 years, human population biology has sought to document and explain processes that have contributed to biological variability in the human species. In particular, it has become clear that the effects of environment upon the genotype are rarely simple, and often not immediately apparent. This has become clearer with the identification of a number of early environmental factors, influencing human growth and development, having long-term biological or behavioural consequences. These include relationships between: (i) the intrauterine environment and adult cardiovascular disease, chronic bronchitis (Barker, 1991) and hypertension (Barker, 1990); (ii) infant diet and cholesterol metabolism (Mott et al., 1982; Hamosh & Hamosh, 1987); (iii) respiratory infection in infancy and chronic lung disease in adult life; and (iv) adverse experiences in childhood and adult psychosocial functioning (Rutter, 1989; Robins & Rutter, 1990; Rutter, 1991). Furthermore, authors have speculated about possible relationships between: (i) the intrauterine environment and non-insulin-dependent diabetes mellitus (Hales & Barker, 1992) and schizophrenia (Murray, Jones & O'Callaghan, 1991); (ii) growth in early childhood and adult immune ageing (Clark et al., 1988, 1989); and (iii) environmental factors influencing adolescent growth and development and ovarian function (Ellison, 1994).

Although often describing disparate phenomena, these observations and speculations are linked by the notion that human developmental processes are environmentally sensitive in a variety of ways, and the outcomes of these processes only become manifest in adult life, or alternatively, appear in childhood and persist into later life. What constitutes an early environmental influence is therefore anything that

25

happens before full developmental maturity is achieved. Of relevance to this type of study is the lifespan perspective of human development, first used to describe processes and states in psychological development (Baltes & Schaie, 1973; Baltes, 1978; Datan & Ginsberg, 1975; Datan & Reese, 1977; Baltes & Brim, 1979, 1980, 1981), but since broadened to encompass all adaptive phenomena in human biology (Leidy, 1994). In this sense, biological anthropologists use the lifespan perspective to evaluate the influence of characteristics at one stage of the life cycle upon subsequent stages (Schell & Beall, 1995). In the present context, the lifespan perspective is limited to the influence of factors during growth and development to different stages in adult life, including ageing and senescence. It should also be appreciated that the human lifespan is much greater than it was even in the recent past, and certainly more than it was in the course of human evolution. Thus it is reasonable to assume that the manifestation of long-term consequences of early environmental influences is greater, partly as a consequence of this.

In this chapter, the human lifespan is set in an evolutionary context, and links are drawn between environmental influences on growth and development, and biological and behavioural characteristics in adult life. These links are viewed from the lifespan perspective, and examples are drawn from various areas including physiology, epidemiology and anthropology.

The human lifespan and human evolution

Although nothing is known about the lifespans of the fossil hominids, some inferences can be made about the evolution of the human lifespan by drawing analogies with extant primate species whose life-history characteristics have been defined (Richard, 1985). On the basis of available DNA sequence homology, humans appear to be most closely related, evolutionarily, with chimpanzees, gorillas, orangutan, with gibbons being the most closely related of the non-ape primates (Harrison et al., 1988) (Figure 3.1). Table 3.1 shows the lifespans of these five primate species, divided into duration of fetal life, time between birth and sexual maturity, and time between sexual maturity and death. Data for non-human primates are from Richard (1985) while for hunter–gatherers, data is for the !Kung bushmen of the Kalahari Desert, and comes from estimates of menarcheal age (Howell, 1976) and life expectancy (Harpending & Jenkins, 1974) parameters. The industrialized population is contemporary British, and the life-history parameters are derived from median age at menarche (Eveleth & Tanner, 1990) and life expectancy at birth (United Nations Children's Fund, 1995).

Excluding industrialized humans, an important feature of human

Table 3.1. *Lifespan variables of living hominoids, in years*

	Fetal lifespan	Birth – sexual maturity	Sexual maturity – death	Total postnatal lifespan
Gibbon	0.6	6	24	30
Orangutan	0.7	11	32	43
Gorilla	0.7	10	32	42
Chimpanzee	0.7	13	30	43
Human (hunter–gatherer)	0.75	17	22	40
Human (industrialized)	0.75	14	60	76

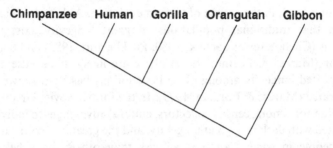

Figure 3.1. Phylogenetic relationships between humans and other primates.

evolution has been the increase in the time between birth and sexual maturity, or of extended childhood. This has been accompanied by a less pronounced increase in longevity, as reflected in the difference in total lifespan between gibbons and apes and humans. There is little difference in longevity between human hunter–gatherers and the apes. However, when longevity of human hunter–gatherers is compared with industrialized humans in the second half of the twentieth century, there is a vast and clear difference between the two. Notably the period between sexual maturity and death is nearly three times longer for industrialized humans than in hunter–gatherers. Thus a striking feature of contemporary human populations is the greatly extended period of adult life. From an evolutionary perspective, the disorders of late onset are a new phenomenon, since they are unlikely to have presented in any proportion in past populations. Furthermore, they are unlikely to have been selected against, since many disorders associated with industrialized populations reveal themselves at postreproductive ages. In this sense, the extended human postreproductive maturity lifespan presents us with a new analytical challenge, since many of the disorders seen are likely to represent genetic–environmental outcomes which are outside of the human design specifications moulded, at

Table 3.2. *Infant mortality rates and life expectancy at birth in 131 countries, according to under-fives mortality rate*

Under-5s mortality[a]	IMR[a] 1960	IMR[a] 1993	LEB[b] 1960	LEB[b] 1993
>170 (n=25)	180	130	37	48
95–170 (n=25)	160	84	41	55
31–94 (n=31)	123	44	50	66
<30 (n=47)	55	11	65	74

[a] per 1000 live births in 1993; [b] years.

population level, in the course of human evolution. Although it has been claimed that traditional populations of great longevity exist in the Caucasus (Chebotarev & Sachuk, 1980; Pitskhelauri, 1982) and highland Ecuador (Mazess & Formin, 1979), these are likely to be false claims. Self-reported longevity among older Ecuadorians has been shown to be exaggerated (Mazess & Forman, 1979). In the former Soviet Union, poor age evaluation among census collectors, cultural advantages to individuals associated with declaring great longevity, and the greater concentration of older people in some Caucasus villages than others, have helped to perpetuate the impression of the existence of long-lived populations (Medvedev, 1986).

Greater population longevity has significance for more than the present industrialized world, since the global trend has been toward increased life expectancy at birth, even in countries with very high levels of under-5s mortality rates (Table 3.2). This may be due largely to reduction in infant mortality rates associated with primary health care, medical intervention and economic development of one sort or another. If this trend continues, greater numbers of people in both the developed and developing world will be susceptible to diseases associated with early environmental influences on human growth and development.

Human growth and development

In relation to other species, the human growth pattern is characterized by a prolonged period of infant dependency, an extended childhood, and a rapid and large acceleration in growth velocity at adolescence leading to physical and sexual maturation (Tanner, 1989). The chronology of the growth and differentiation of specific organs and tissues varies, and it has been suggested that there may be critical periods when environmental factors such as nutritional stress or exposure to infection can significantly alter the developmental process from its genetic trajectory. McCance (1962) identified the importance of chronological time, or age, on the pattern of development in relation to nutritional stress in humans and other mammalian species. Since that time, work on the developmental biology of invertebrate species has revealed various and subtle ways in which environmental changes or cues at different stages of development can influence morphology (Robertson, 1963; Bryant, French & Bryant, 1981; Bryant & Simpson, 1984; French, 1989). Following the rich literature on this topic, Lucas (1991) used the term 'programming' to describe such phenomena when associated with some health-related outcome.

Lucas (1991) distinguishes two ways in which programming could happen: (i) by the induction, deletion, or impaired development of a permanent somatic structure as the result of a stimulus or insult operating at a critical time; and (ii) by physiological 'setting' by an early stimulus or insult at a 'sensitive' period, resulting in long-term consequences for function, the effects being either immediate or deferred. For the purposes of subsequent discussion, category (i) phenomena are called 'distorted morphology', while category (ii) phenomena are referred to as 'metabolic imprinting'.

Lucas (1991) does not discuss the position of behaviour in his classification. Although ambiguous, certain aspects of behavioural development can be easily classified. For example, environmentally modified brain growth and development during any critical period of development could have behavioural correlates (Smart, 1991), and would be included in category (i) phenomena; early events in a child's life (including attachment to parents in the first minutes and hours after birth) could influence the quality of all future bonds to other individuals (Klaus & Kennell, 1976), and falls into category (ii).

There are some problems associated with the use of the early environment-later outcome framework. First, the definition of an early environment is not straightforward. It may be argued that 'early environment' is intrauterine, or ceases by the age of one year, on the basis of the observation that most critical periods in development may have been passed by this age. However, while this may be true for the development of physiological systems which are associated with cardiovascular disease risk (Barker, this volume), it is not true of critical periods in behavioural development. Behavioural development is closely related to physical and physiological growth and development, set in social contexts (Lerner, 1984). There are various pathways from childhood experiences to adult psychosocial functioning, but a common feature is that long-term effects depend for their occurrence on several, sometimes many, intervening links in a chain of indirect connections (Rutter, 1991). When such links are all present, these long-term effects can be strong, but in their absence, there may be no enduring consequences of even severe early adversities (Rutter, 1991). However, epidemiological associations have been found between stress in early childhood and adult behaviour and parenting (Wadsworth, 1984), and between parents' divorce and children's experience of depression, particularly if divorce took place before the child was of school age (Wadsworth & Maclean, 1986).

In addition, physiological processes which take place across childhood and which result in modified morphology may have long-term consequences of the health and human biology of the adult. An example of this is the phenotypic plasticity of muscular development in relation to nutritional stress across childhood. The definition of early environment should therefore include the social environment, and extend across the entire course of growth and development. Thus environmental influences of any kind, at any stage of growth and development, and which may have lasting effects on the adult, or may only come to be expressed in later adult life, fall into this framework.

Long-term consequences of distorted morphology

Three examples of long-term consequences of possible distorted morphology are: (i) fetal malnutrition and adult non-insulin-dependent diabetes mellitus; (ii) viral infection in pregnancy, fetal brain development and schizophrenia; and (iii) childhood undernutrition and efficiency of muscular contraction. Although the first two examples illustrate negative outcomes, the third illustrates a positive outcome.

An epidemiological relationship between non-insulin-dependent diabetes mellitus (NIDDM) or glucose tolerance and low birthweight or weight at one year of age (Hales & Barker, 1992) led to the proposal of a mechanism relating impaired development of the pancreas in the undernourished fetus to the development of NIDDM in adults (Hales & Barker, 1992). This is in opposition to the genetic explanation first put forward by Neel (1962). The postulated mechanism is illustrated in Figure 3.2. Briefly, the authors' speculation is that maternal malnutrition and/or other maternal placental abnormalities lead to fetal malnutrition. The consequences of a generalized undernutrition include reduced fetal growth, and a postulated reduced Beta cell mass and islet function of the pancreas. Deficiencies of protein and amino acids are believed to be of particular importance in creating these defects of structure and function. Citing Swenne (1992) and De Gasparo et al. (1978), the authors put forward the view that amino acids are the major factors controlling beta cell growth and development, and insulin secretion until late gestation. They also suggest

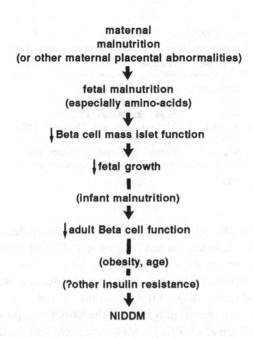

**maternal
malnutrition
(or other maternal placental abnormalities)**

↓

**fetal malnutrition
(especially amino-acids)**

↓

↓**Beta cell mass islet function**

↓

↓**fetal growth**

❚

(infant malnutrition)

↓

↓**adult Beta cell function**

❚

(obesity, age)

❚

(?other insulin resistance)

↓

NIDDM

Figure 3.2. Postulated mechanisms underpinning the 'thrifty phenotype' hypothesis of Hales and Barker (1992).

that impaired beta-cell function persists into adulthood, predisposing the individual to NIDDM. This disease then expresses itself in association with environmental risk factors, obesity and physical inactivity in particular, as well as increasing age, and possibly other processes leading to insulin resistance.

Recent evidence suggests that the increased prevalence of diabetes in adults who were growth retarded during fetal life is mediated by increased insulin resistance, rather than by abnormalities of insulin secretion (Philips *et al.*, 1994*a,b*). However, genetic explanations appear to be ascendant again despite many obstacles hindering the study of this field (McCarthy *et al.*, 1994), with claims of a candidate gene for NIDDM (De Fronzo, Bonadonna & Ferranini, 1992). Furthermore, U-shaped relationships between the prevalence of NIDDM and birthweight among Pima Indians in Arizona have lead to the suggestion that high mortality of low birthweight infants may be associated with selective survival in infancy of individuals who are genetically predisposed to insulin resistance and diabetes (McCance *et al.*, 1994).

The second example is that of schizophrenia. There is clear seasonality in the birth-rates of schizophrenics, with greater numbers being born in late

Table 3.3. *Decline in schizophrenia incidence*

	Period	Decrease	Authors
England & Wales	1965–86	about 50%	Der, Gupta & Murray, 1990
Scotland	1970–81	48%	Dickson & Kendall, 1986
Scotland	1969–78	40%	Eagles & Whalley, 1985
Denmark	1970–84	37–44%	Munk-Jorgensen & Jorgensen, 1986
New Zealand	1974–84	37%	Joyce, 1987
Australia	1967–77	9%	Parker, O'Donnell & Walter, 1985

From Der, Gupta & Murray, 1990

winter or early spring (Bradbury & Miller, 1985). Furthermore, a decline in the treated incidence of schizophrenia has been observed in a number of industrialized nations in recent decades (Table 3.3). The postulated link between these two observations is that of maternal infection with a neuropathogen, the influenza virus. Viral infection of the mother is believed to adversely affect the developing brain of the fetus, predisposing it to later schizophrenia (Sham et al.,1992), a view consistent with both the neurodevelopmental (Castle, 1993) and genetic (McGuffin & Stuart, 1986) theories of schizophrenia.

The association between influenza and schizophrenia was first alluded to by Menninger (1928) who described 67 cases of dementia praecox in a group of 175 patients with mental illness following the 1918 European influenza pandemic. The winter birth excess in schizophrenics has been associated with higher than expected rates of influenza in the sixth month of gestation, in the Danish population between 1911 and 1950 (Barr, Mednick & Munk-Jorgensen, 1990). Furthermore, Mednick et al. (1988) found that individuals in utero during the second trimester of pregnancy during the 1957 influenza epidemic in Denmark had a higher than expected susceptibility to subsequent schizophrenia, while Sham et al. (1992) showed that influenza epidemics were consistently correlated with higher rates of subsequent schizophrenic births in England and Wales between 1939 and 1960. The reported decline in the incidence of schizophrenia in Britain, Denmark, Australia and New Zealand has been attributed, at least in part, to general improvements in living conditions in recent decades (Sham et al., 1992), resulting in a decline in influenza morbidity. However, epidemiological studies do not explain why only some mothers exposed to influenza during the second trimester of pregnancy produce offspring who develop schizophrenia. It has been proposed that this pattern of schizophrenia could be initiated by autoimmune disorders arising in genetically at-risk mothers and fetuses exposed to a viral trigger (Wright, Gill & Murray, 1993).

Since the influenza virus does not cross the placenta (O'Callaghan *et al.*, 1991), it has been suggested that the process is most likely to be mediated immunologically, possibly by the induction of maternal antibodies by the influenza virus, which unlike the virus, are able to cross the placenta and immature blood–brain barrier and cross-react with fetal tissues, interfering with neuronal migration (Wright *et al.*, 1993). The cytotoxic response to influenza virus depends on Class I human leucocyte antigens (HLA), while antibody response depends on Class II HLA (Brostoff *et al.*, 1991). Paranoid schizophrenia has been associated with HLA Class I A9 (McGuffin & Stuart, 1986; Owen & McGuffin, 1991), and HLA Class II DR8 in Japanese schizophrenics (Miyanaga, Machiyama & Jugi, 1984) and the Class II HLA DRw6 in African Americans (Rabin *et al.*, 1987). Thus the extent of the potentially neuronally damaging immune response may be influenced by genetic variation in HLA both within- and between-populations, and that patterns of schizophrenia morbidity may have some basis in the impaired development of permanent somatic structures in the brain of fetuses of genetically susceptible mothers (Wright *et al.*, 1993), and can be regarded as an outcome of distorted morphology.

The third example of postulated distorted morphology with possible long-term consequences is that of childhood undernutrition and the efficiency of muscular contraction. Mechanical efficiency is the ratio between work done and the energy expended in doing it, and Shetty (1993) has summarized reports of this variable in adults experiencing chronic energy deficiency, compared with well-nourished controls (Table 3.4).

Although the comparison of these studies, which were carried out under a variety of conditions with disparate experimental designs, is not straightforward (Shetty, 1993), in general, undernourished subjects appear to use less energy in performing some standardized task than do their well-nourished controls. Waterlow (1990) proposed a physiological explanation for this phenomenon, relating undernutrition, the selective development of slow and fast twitch muscle fibres, their relative energetic efficiencies in muscular contraction, and overall work performance.

The relation between the force developed or the amount of mechanical work done to the amount of ATP used is called the contraction coupling efficiency (Waterlow, 1990). Slow twitch fibres have a higher contraction-coupling efficiency than fast twitch fibres (Awan & Goldspink, 1972; Crow & Kushmerik, 1982). Muscles with a high proportion of slow twitch fibres use less ATP per unit of isometric tension than muscles containing a high proportion of fast twitch fibres (Wendt & Gibbs, 1973; Goldspink, 1975). In conditions of low energy intake and in hypothyroidism there is a reduction in the proportion and diameter of fast twitch fibres (Russell *et al.*,

Table 3.4. *Mechanical efficiency in chronic undernutrition*

Country	Mechanical efficiency			Author
	Gross	Net	Delta	
Stepping				
Jamaica	Increased			Ashworth, 1968
India		Increased		Kulkarni & Shetty, 1992
Bicycle ergometer				
Java	Increased			Edmundson, 1979
Brazil	No change			Desai *et al.*, 1984
India	Increased		Decreased	Satyanarayana, Venkataramana & Rao, 1989
Treadmill				
Colombia	Increased	Increased	No change	Spurr *et al.*, 1984
Gambia		Increased	Increased	Minghelli *et al.*, 1990
India		No change	Decreased	Shetty, 1993

After Shetty, 1993.

1984; Wiles *et al.*, 1979), with increased contraction coupling efficiency in hypothyroidism (Wiles *et al.*, 1979).

Studies of low intensity, long duration training in healthy adults showed a reduction in fast twitch fibres in the triceps brachii and quadriceps muscles, while slow twitch fibres were not affected (Schantz, Henriksson & Jansson, 1983). Similarly, investigations of muscle size and composition in malnourished patients showed that the size of the slow twitch fibres in the calf muscle was better preserved than that of the fast twitch fibres. In the developing world, low dietary energy availability and prolonged low intensity work output may lead to such changes in muscular development, with the associated lower energy expenditure made possible by them.

Differential development of slow twitch over fast twitch fibres in the course of the growth faltering has not been demonstrated. However, muscle phenotype characteristics may be modified or determined by thyroid hormone status (Nicol & Bruce, 1981; Jollesz & Sreter, 1981). Notably, clinically hypothyroid patients show significantly lower energy expenditure compared with euthyroid controls during quadriceps muscle function tests (Wiles *et al.*, 1979). A possible mechanism for this effect might be that thyroid hormones influence the sequestration and release of calcium ions by the sarcoplasmic reticulum, and ATP consumption in calcium pumping during the contraction cycle, thereby increasing overall efficiency (Suko, 1973; Limas, 1978, Hardeveld & Clausen, 1984). It is not clear whether the disappearance of the low thyroid status associated with growth faltering leads to reversion of the muscle phenotype to one with a

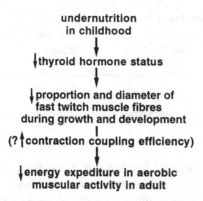

undernutrition
in childhood
↓
↓thyroid hormone status
↓
↓ proportion and diameter of
fast twitch muscle fibres
during growth and development
↓
(? ↑contraction coupling efficiency)
↓
↓energy expediture in aerobic
muscular activity in adult

Figure 3.3. Postulated mechanism whereby undernutrition in childhood may influence the efficiency of muscular contraction and energy cost of physical activity in adulthood (from Waterlow, 1990).

greater proportion of fast twitch fibres. If nutritional stress persists through life, low T3 will persist, and with it, greater muscular efficiency associated with selective persistence of slow twitch fibres over fast twitch fibres. A postulated mechanism whereby this type of distorted morphology arising out of undernutrition in childhood could have a positive outcome for physiological function is illustrated in Figure 3.3.

Long-term consequences of metabolic imprinting

Two examples of possible long-term consequences of metabolic imprinting are given: (i) infant feeding and cholesterol metabolism; and (ii) the timing of onset of puberty and fecundity and breast cancer. Both of these speculations are based on the knowledge that the hormonal imprinting of a variety of metabolic processes takes place in the course of growth and development, and that such physiological setting can be modified at critical periods by exogenous factors.

The first example is based on speculations that the cholesterol intake from breast milk has an important role in the development of cholesterol regulatory mechanisms, and that preweaning cholesterol intake may prevent diet-induced hypercholesterolaemia later in life (Fomon, 1971; Reiser & Sidelman, 1972). There are various strands of evidence that support this view. Serum cholesterol of infants differs according to their diet (Fomon & Bartels, 1960), while children in the United States fed a low cholesterol formula during infancy had a lower mean serum cholesterol value between the ages of 7 and 12 years than children fed greater amounts of cholesterol in infancy (Hodgson *et al.*, 1976). Marmot and colleagues

Table 3.5. *Menarche and ovarian cycle length*

	Age at menarche (years)	Median OCL (days)	Range (days)
Gainj (PNG)	18.4	36.0	32.8–39.8
USA	12.8	26.1	23.1–29.3

Johnson, Wood, Campbell & Maslar, 1987.

(1980) found that women who had been breast fed had significantly lower mean plasma cholesterol than women who had been bottle fed.

However, studies with experimental animals do not provide unequivocal support for these views (Hamosh & Hamosh, 1987; Hamosh, 1988), although Mott *et al.* (1991) have proposed that metabolic imprinting by breast feeding may result in more efficient utilization of cholesterol by breast-fed baboons, including both greater cholesterol production rates and absorption, than in formula-fed individuals. This might be advantageous during infancy, when cholesterol is a major requirement for the formation of new cell membranes, but disadvantageous in later life, if exposed to a high cholesterol, high fat diet which raises circulating serum lipoprotein levels and accelerates atherogenesis. They speculate that numerous growth factors and hormones in breast milk might mediate this process. However, at present, this speculation is too vaguely formulated to be testable.

The second example involves speculations about relationships between the onset of puberty and female fecundity. Luteal function varies markedly between populations (Ellison *et al.*, 1993). In the developing world, biological regulation of ovarian function may include such factors as lactation, nutritional state, energy balance, physical activity levels and stress (Rosetta, 1990). Poor nutritional status can also influence the onset and timing of puberty, and it is possible that the timing of puberty and subsequent ovulatory function in adult life may be linked. Data shown in Table 3.5 on median age at menarche and ovarian cycle length for two groups, the Gainj of Papua New Guinea, and a sample of women in the United States, are suggestive. The Gainj are believed to experience considerable nutritional stress and have both extremely late median age at menarche and long ovarian cycle length, while the United States sample experiences little or no nutritional stress and has early menarche and short cycle length (Johnson *et al.*, 1987).

Support for the link between developmental processes in adolescence and ovarian function is sparse. However, there is evidence of higher frequencies of oligomenorrhoea and dysmenorrhoea among late maturers than early maturers when adult, in a unique sample of North American women (Gardner, 1983; Gardner & Valadian, 1983). Ellison (1994) has suggested that chronic differences in adult gonadal function are either: (i) a consequence of developmental processes operating during childhood and adolescence to establish adult set-points of the hypothalamic–pituitary–gonadal axis; or (ii) that the differences may result from acute ecological factors. It is unlikely, however, that these two conjectures are mutually exclusive. Indeed, in the developing world, such ecological factors as poor nutritional state may serve to create different, lower endocrine set-points than would be the case if those stresses were absent. In this way, nutritional stress in adolescence could operate in the metabolic imprinting of the reproductive system.

Although it is not clear whether altered hormonal set-points are likely to result in modified ovarian function, endogenous levels of reproductive steroids are known to be associated with breast, ovarian and endometrial cancer risk (Pike, 1985). Furthermore, metabolic imprinting of males in adolescence cannot be dismissed, and elevated levels of testosterone and its metabolite dihydrotestosterone have been implicated in the aetiology of prostate cancer (Pike, 1985). Thus aspects of metabolic imprinting in adolescence may have long-term consequences with respect to cancer morbidity.

Conclusions

A wide range of phenomena in which early environmental factors influencing human growth and development have long-term biological and behavioural consequences have been described by workers in a diversity of disciplines. One reason why they are becoming more apparent is that the extended period of adult life experienced by industrialized populations in the twentieth century allows the greater expression of degenerative diseases in populations that are living longer, beyond the lifespans which may have been typical of populations from which humans evolved. The developmental lifespan perspective is appropriate to the study of such phenomena, and within this framework, most long-term outcomes of early environmental influences on growth and development fall into one of two categories: (i) distorted morphology; and (ii) metabolic imprinting. The former consists of phenomena which may arise from impaired development of a somatic structure, while the second comprises phenomena which may be the outcome of physiological 'setting' of some mechanism, probably hormonal, at some critical time in growth and development. Although most of the evidence is at best suggestive or associational, category (i) phenomena are far less speculative and more common than category (ii) phenomena.

References

Ashworth, A. (1968). An investigation of very low caloric intakes reported in Jamaica. *British Journal of Nutrition*, **22**, 341–55.

Awan, M.Z. & Goldspink, G. (1972). Energetics of the development and maintenance of isometric tension by mammalian fast and slow muscles. *Journal of Mechanochemical Cell Motility*, **1**, 97–108.

Baltes, P.B. (ed) (1978). *Life-span Development and Behavior* (Vol. 1). New York: Academic Press.

Baltes, P.B. & Brim, O.G., Jr. (eds) (1979). *Life-span Development and Behavior* (Vol. 2). New York: Academic Press.

Baltes, P.B. & Brim, O.G., Jr. (eds) (1980). *Life-span Development and Behavior* (Vol. 3). New York: Academic Press.

Baltes, P.B. & Brim, O.G., Jr. (eds) (1981). *Life-span Development and Behavior* (Vol. 4). New York: Academic Press.

Baltes, P.B. & Schaie, K.W. (1973). *Life-span Developmental Psychology: Personality and Socialization*. New York: Academic Press.

Barker, D.J.P. (1990). The intrauterine origins of adult hypertension. In *Fetal Autonomy and Adaptation*, ed. G.S. Dawes, pp. 42–57. Chichester, West Sussex: John Wiley and Sons.

Barker, D.J.P. (1991). The intrauterine environment and adult cardiovascular disease. In *The Childhood Environment and Adult Disease*, ed. G.R. Bock and J. Whelan, pp. 3–10. Chichester, West Sussex: John Wiley and Sons.

Barr, C.E., Mednick, S.A. & Monk-Jorgensen, P. (1990). Exposure to influenza epidemics during gestation and adult schizophrenia. *Archives of General Psychiatry*, **47**, 869–74.

Bradbury, T.N. & Miller, G.A. (1985). Season of birth in schizophrenia: a review of evidence, methodology, and aetiology. *Psychology Bulletin*, **98**, 569–74.

Brostoff, J., Scadding, G.K., Male, D.K. & Roitt, I.M. (1991). Introduction to immune responses. In *Clinical Immunology*, ed. J. Brostoff, G.K. Scadding, D.K. Male and I.M. Roitt, pp. 1–1.7. London: Gower Medical Publishing.

Bryant, S.V., French, V. & Bryant, P.J. (1981). Distal regeneration and symmetry. *Science*, **212**, 993–1002.

Bryant, S.V. & Simpson, P. (1984). Intrinsic and extrinsic control of growth in developing organs. *Quarterly Review of Biology*, **59**, 387–415.

Castle, D.J. (1993). Some current controversies in the epidemiology of schizophrenia. Current medical literature: *Psychiatry*, **4**, 3–7.

Chebotarev, D.F. & Sachuk, N.N. (1980). Aging. Union of Soviet Socialist Republics. In *International Handbook on Ageing*, ed. E. Palmore, pp. 401–17. London: The Macmillan Press.

Clark, G.A., Aldwin, C.M., Hall, N.R., Spiro, A. & Goldstein, A. (1989). Is poor early growth related to adult immune aging? A follow-up study. *American Journal of Human Biology*, **1**, 331–7.

Clark, G.A., Hall, N.R., Aldwin, C.M., Harris, J.M., Borkan, G.A. & Srinivasan, M. (1988). Measures of poor early growth are correlated with lower adult levels of thymosin-1. *Human Biology*, **60**, 435–51.

Crow, M.T. & Kushmerick, M.J. (1982). Chemical energetics of mammalian muscle. *Journal of General Physiology*, **79**, 147–66.

Datan, N. & Ginsberg, L.H. (1975). *Life-span Developmental Psychology: Normative Life Crises*. New York: Academic Press.

Datan, N. & Reese, H.W. (1977). *Life-span Developmental Psychology: Dialectical Perspectives on Experimental Psychology.* New York: Academic Press.

Desai, I.D., Wadell, C., Ditra, S., Dutra de Oliviera, S., Duarte Robazzi, M.L., Cevallos Romero, L.S., Desai, M.I., Vichi, F.L., Bradfield, R.B. and Dutra de Oliviera, J.E. (1984). Marginal malnutrition and reduced physical work capacity of migrant adolescent boys in Southern Brazil. *American Journal of Clinical Nutrition*, **40**, 135–45.

Der, G., Gupta, S. & Murray, R.M. (1990). Is schizophrenia disappearing? *Lancet*, **335**, 513–16.

Dickson, W.E. & Kendall, R.E. (1986). Does maintenance lithium therapy prevent recurrences of mania under ordinary clinical conditions? *Psychological Medicine*, **16**, 521–30.

Eagles, J.M. & Whalley, L.J. (1985). Decline in the diagnosis of schizophrenia among first admissions to Scottish mental hospitals from 1969–1978. *British Journal of Psychiatry*, **146**, 151–4.

Edmundson, W. (1979). Individual variations in basal metabolic rate and mechanical work efficiency in East Java. *Ecology of Food and Nutrition*, **8**, 189–95.

Ellison, P.T. (1994). Developmental influences on adult reproductive function. *American Journal of Physical Anthropology*, Supplement, **18**, 85.

Ellison, P.T., Lipson, S.F., O'Rourke, M.T., Bentley, G.R., Harrigan, A.M., Panter-Brick, C. & Vitzthum, V.J. (1993). Population variation in ovarian function. *Lancet*, **342**, 433–4.

Eveleth, P.B. & Tanner, J.M. (1990). *Worldwide variation in human growth.* Cambridge: Cambridge University Press.

Fomon, S.J. (1971). A pediatrician looks at early nutrition. Bulletin of the *New York Academy of Medicine*, **47**, 569–78.

Fomon, S.J. & Bartels, D.J. (1960). Concentrations of cholesterol in serum of infants in relation to diet. *Journal of Diseases in Childhood*, **99**, 43–6.

French, V. (1989). The control of growth and size during development. In *The Physiology of Human Growth*, ed. J.M. Tanner and M.A. Preece, pp. 11–28. Cambridge: Cambridge University Press.

De Fronzo, R.A., Bondonna, R.C. & Ferrannini, E. (1992). Pathogenesis of NIDDM. *Diabetes Care*, **15**, 318–58.

Gardner, J. (1983). Adolescent menstrual characteristics as predictors of gynaecological health. *Annals of Human Biology*, **10**, 31–40.

Gardner, J. & Valadian, I. (1983). Changes over thirty years in an index of gynaecological health. *Annals of Human Biology*, **10**, 41–55.

De Gasparo, M., Milner, G.R., Norris, P.D., Milner, R.D.G. (1978). Effect of glucose and amino acids on foetal rat pancreatic growth and insulin secretion in vitro. *Journal of Endocrinology*, **77**, 241–8.

Goldspink, G. (1975). Biochemical energetics for fast and slow muscles. In: *Comparative Physiology – Functional Aspects of Structural Materials*, ed. L. Bolis, H.P. Maddrell and K. Schmidt-Nielsen, pp. 173–85. Amsterdam: North Holland Publishing Company.

Hales, C.N. & Barker, D.J.P. (1992). Type 2 (non-insulin-dependent) diabetes mellitus: the thrifty phenotype hypothesis. *Diabetologia*, **35**, 595–601.

Hamosh, M. (1988). Does infant nutrition affect adiposity and cholesterol levels in the adult? *Journal of Pediatric Gastroenterology and Nutrition*, **7**, 10–16.

Hamosh, M. & Hamosh, P. (1987). Does nutrition in early life have long-term metabolic effects? Can animal models be used to predict these effects in the human? *In Human Lactation 3: the Effects of Human Milk on the Recipient Infant*, ed. A.S. Goldman, A.S. Atkinson and L.A. Hanson, pp. 37–55. New York: Plenum Publishing Corporation.

Hardeveld, C. van, & Clausen, T. (1984). Effect of thyroid status on K + -stimulated metabolism and 45Ca exchange in rat skeletal muscle. *American Journal of Physiology*, **247**, E421–30.

Harpending, H.C. & Jenkins, T. (1974). !Kung population structure. In *Genetic Distance*, ed. J.F. Crow and C. Dennison, pp. 137–65. New York: Plenum Press.

Harrison, G.A., Pilbeam, D.R., Tanner, J.M. & Baker, P.T. (1988). *Human Biology. An Introduction to Human Variation, Evolution, Growth and Adaptability*. Oxford: Oxford University Press.

Hodgson, P.A., Ellefson, R.D., Elveback, L.R., Harris, L.E., Nelson, R.A. & Weidman, W.H. (1976). Comparison of serum cholesterol in children fed high, moderate or low cholesterol milk diets during neonatal period. *Metabolism*, **25**, 739–46.

Howell, N. (1976). The population of the Dobe Area !Kung. In *Kalahari Hunter-Gatherers. Studies of the !Kung San and Their Neighbours*, ed. R.B. Lee and I. DeVore, pp. 137–51. Cambridge, Massachusetts: Harvard University Press.

Johnson, P.L., Wood, J.W., Campbell, K.L. & Maslar, I.A. (1987). Long ovarian cycles in women of Highland New Guinea. *Human Biology*, **59**, 837–45.

Jollesz, F. & Sreter, F.A. (1981). Development, innervation, and activity-pattern induced changes in skeletal muscle. *Annual Reviews in Physiology*, **43**, 531–52.

Joyce, P.R. (1987). Changing trends in first admissions and readmissions for mania and schizophrenia in New Zealand. *Australia and New Zealand Journal for Psychiatry*, **21**, 82–6.

Klaus, M. & Kennell, J. (1976). *Maternal–Infant Bonding*. St Louis: Mosby.

Kulkarni, R.N. & Shetty, P.S. (1992). Net mechanical efficiency during stepping in chronically energy deficient human subjects. *Annals of Human Biology*, **19**, 421–5.

Leidy, L.E. (1994). The lifespan approach: but isn't it too broad? *American Journal of Physical Anthropology*, Supplement, **18**, 128.

Lerner, R.M. (1984). *Individual and Context in Developmental Psychology: Conceptual and Empirical*. New York: Academic Press.

Limas, C.J. (1978). Calcium transport ATPase of cardiac sarcoplasmic reticulum in experimental hyperthyroidism. *American Journal of Physiology*, **235**, H745–51.

Lucas, A. (1991). Programming by early nutrition in man. In *The Childhood Environment and Adult Disease*, ed. G.R. Bock & J. Whelan, pp. 38–50. Chichester, West Sussex: John Wiley and Sons Ltd.

McCance, R.A. (1962). Food, growth, and time. *Lancet*, **ii**, 621–6.

McCance, D.R., Pettitt, D.J., Hanson, R.L., Jacobsson, L.T.H., Knowler, W.C. & Bennett, P.H. (1994). Birth weight and non-insulin dependent diabetes: thrifty genotype, thrifty phenotype, or surviving small baby genotype? *British Medical Journal*, **308**, 942–5.

McCarthy, M.I., Froguel, P. & Hitman, G.A. (1994). The genetics of non-insulin-dependent diabetes mellitus: tools and aims. *Diabetologia*, **37**, 959–68.

McGuffin, P. & Stuart, E. (1986). Genetic markers in schizophrenia. *Human Heredity*, **36**, 65–88.

Marmot, M.G., Page, C.M., Atkins, E. & Douglas, J.W.B. (1980). Effect of breast feeding on plasma cholesterol and weight in young adults. *Journal of Epidemiology and Community Health*, **34**, 164–7.

Martyn, C.N. (1991). Childhood infection and adult disease. In *The Childhood Environment and Adult Disease*, ed. G.R. Bock & J. Whelan, pp. 93–102. Chichester, West Sussex: John Wiley and Sons Ltd.

Mazess, R.B. & Forman, S.H. (1979). Longevity and age exaggeration in Vilcabamba, Ecuador. *Journal of Gerontology*, **34**, 94–8.

Mednick, S.A., Machon, R.A., Huttenen, M.O. & Bonet, D. (1988). Adult schizophrenia following prenatal exposure to an influenza epidemic. *Archives of General Psychiatry*, **45**, 189–92.

Medvedev, Z.A. (1986). Age structure of Soviet population in the Caucasus: facts and myths. In *The Biology of Human Ageing*, ed. A.H. Bittles and K.J. Collins, pp. 181–99. Cambridge: Cambridge University Press.

Menninger, K.A. (1928). The schizophrenic syndromes as a product of acute infectious disease. *Archives of Neurological Psychiatry*, **20**, 464–81.

Minghelli, G., Schutz, Y., Charbonnier, A., Whitehead, R. & Jequier, E. (1990). Reduced 24 hour energy expenditure and basal metabolic rate in Gambian men. *American Journal of Clinical Nutrition*, **51**, 563–70.

Miyanaga, K., Machiyama, Y. & Juji, T. (1984). Schizophrenic disorders and HLA DR antigens. *Biological Psychiatry*, **19**, 121–9.

Mott, G.E., Lewis, D.S. & McGill, H.C. (1991). Programming of cholesterol metabolism by breast or formula feeding. In *The Childhood Environment and Adult Disease*, ed. G.R. Bock & J. Whelan, pp. 56–66. Chichester, West Sussex: John Wiley and Sons Ltd.

Mott, G.E., McMahan, C.A., Kelley, J.L., Farley, C. M. & Mcgill, H.C. (1982). Influence of infant and juvenile diets on serum cholesterol, lipoprotein cholesterol, and apolipoprotein concentrations in juvenile baboons. *Atherosclerosis*, **45**, 191–202.

Munk-Jorgensen, P. & Jorgensen, P. (1986). Decreasing rates of first admission diagnoses of schizophrenia among females in Denmark 1970–1984. *Acta Psychiatrica Scandinavica*, **74**, 379–83.

Murray, R.M., Jones, P. & O'Callaghan, E. (1991). Fetal brain development and later schizophrenia. In *The Childhood Environment and Adult Disease*, ed. G.R. Bock & J. Whelan, pp. 155–63. Chichester, West Sussex: John Wiley and Sons Ltd.

Neel, J.V. (1962). Diabetes Mellitus: a 'thrifty' genotype rendered detrimental by 'progress'? *Journal of Human Genetics*, **14**, 353–62.

Nicol, C.J.M. & Bruce, D.S. (1981). Effect of hyperthyroidism on the contractile and histochemical properties of fast and slow twitch skeletal muscle in the rat. *Pflugers Archiv*, **390**, 73–9.

O'Callaghan, E., Sham, P., Takei, N., Glover, G. & Murray, R.M. (1991). Schizophrenia after prenatal exposure to 1957 A2 influenza epidemic. *Lancet*, **337**, 1248–50.

Owen, M.J. & McGuffin, P. (1991). DNA and classical genetic markers in schizophrenia. *European Archives in Psychiatry and Clinical Neuroscience*, **240**, 197–203.

Parker, G., O'Donnell, M. & Walter, S. (1985). Changes in the diagnosis of the functional psychoses associated with the introduction of lithium. *British Journal of Psychiatry*, **146**, 377–82.

Phillips, D.I.W., Hirst, S., Clark, P.M.S., Hales, C.N. & Osmond, C. (1994b). Fetal growth and insulin secretion in adult life. *Diabetologia*, **37**, 592–6.

Phillips, D.I.W., Barker, D.J.P., Hales, C.N., Hirst, S. & Osmond, C. (1994b). Thinness at birth and insulin resistance in adult life. *Diabetologia*, **37**, 150–4.

Pike, M. (1985). Endogenous hormones. In *Cancer. Risks and Prevention*, ed. M.P. Vessey and M. Gray, pp. 195–210.

Pitskhelauri, G.Z. (1982). *The Longliving of Soviet Georgia*. New York: Human Science Press.

Reiser, R. & Sidelman, Z. (1972). Control of serum cholesterol homeostasis by cholesterol in the milk of the suckling rat. *Journal of Nutrition*, **102**, 1009–16.

Richards, A.F. (1985). *Primates in Nature*. New York: W.H. Freeman and Company.

Robertson, F.W. (1963). The ecological genetics of growth in Drosophila. 6. The genetic correlation between the duration of the larval period and body size in relation to larval diet. *Genetical Research Cambridge*, **4**, 74–92.

Robins, L. & Rutter, M. (1990). *Straight and Devious Pathways from Childhood to Adulthood*. Cambridge: Cambridge University Press.

Rosetta, L. (1990). Biological aspects of fertility among third world populations. In *Fertility and Resources*, ed. V. Reynolds and J. Landers, pp. 18–34. Cambridge: Cambridge University Press.

Russell, D.McR., Walker, P.M., Leiter, L.A., Sima, A.A.F., Tanner, W.K., Mickle, D.A.G., Whitwell, J., Marliss, E.B. & Jeejeebhoy, K.N. (1984). Metabolic and structural changes in skeletal muscle during hypocaloric dieting. *American Journal of Clinical Nutrition*, **39**, 503–13.

Rutter, M. (1989). Pathways from childhood to adult life. *Journal of Child Psychiatry and Allied Disciplines*, **30**, 23–51.

Rutter, M. (1991). Childhood experiences and adult psychosocial functioning. In *The Childhood Environment and Adult Disease*, ed. G.R. Bock, J. Whelan, pp. 189–200. Chichester: John Wiley & Sons Ltd.

Satyanarayana, K., Venkataramana, Y. & Rao, S.M. (1989). *Nutrition and work performance: studies carried out in India. Proceedings of XIVth International Congress of Nutrition*, pp. 98–9. Seoul, Korea: Korean Nutrition Society.

Schantz, P., Henriksson, J. & Jansson, E. (1983). Adaptation of human skeletal muscle to endurance training of long duration. *Clinical Physiology*, **3**, 141–51.

Schell, L.M. (1995). Human biological adaptability with special emphasis on plasticity: history, development and problems for future research. In *Plasticity*, ed. B. Bogin, C.G.N. Mascie-Taylor and G.A. Harrison, pp. 213–37. Cambridge: Cambridge University Press.

Sham, P.C., O'Callaghan, E., Takei, N. *et al.* (1992). Schizophrenia following prenatal exposure to influenza epidemics between 1939 and 1960. *British Journal of Psychiatry*, **160**, 461–6.

Shetty, P.S. (1993). Chronic undernutrition and metabolic adaptation. *Proceedings of the Nutrition Society*, **52**, 267–84.

Smart, J.L. (1991). Critical periods in brain development. In *Childhood Environment and Adult Disease*, ed. G.R. Bock and J. Whelan, pp. 109–124. Chichester: John Wiley & Sons Ltd.

Spurr, G.B., Barac-Nieto, M., Reina, J.C. & Ramirez, R. (1984). Marginal malnutrition in school-aged Colombian boys: efficiency of treadmill walking in submaximal exercise. *American Journal of Clinical Nutrition*, **36**, 452–9.

Suko, J. (1973). The calcium pump of cardiac sarcoplasmic reticulum. Functional alterations at different levels of thyroid state in rabbits. *Journal of Physiology*, **228**, 563–82.

Swenne, I. (1992). Pancreatic Beta-cell growth and diabetes mellitus. *Diabetologia*, **35**, 193–201.

Tanner, J.M. (1989). *Foetus into Man*. Ware, Hertfordshire: Castlemead Publications.

United Nations Children's Fund (1995). *The State of the World's Children 1995*. Oxford: Oxford University Press.

Wadsworth, M.E.J. (1984). Early stress and association with adult behaviour and parenting. In *Stress and Disability in Childhood*, ed. N.R. Butler and B.D. Corner, pp. 100. Bristol: John Wright.

Wadsworth, M.E.J. & Maclean, M. (1986). Parents' divorce and children's life chances. *Children and Youth Services Review*, **8**, 145–59.

Waterlow, J.C. (1990). Mechanisms of adaptation to low energy intakes. In *Diet and Disease in Traditional and Developing Societies*, ed. G.A. Harrison & J.C. Waterlow, pp. 5–23. Cambridge: Cambridge University Press.

Wendt, I.R. & Gibbs, C.L. (1973). Energy production of rat extensor digitorum longus muscle. *American Journal of Physiology*, **224**, 1081–6.

Wiles, C.M., Young, A., Jones, D.A. & Edwards, R.H.T. (1979). Muscle relaxation rate, fibre-type composition and energy turnover in hyper- and hypo- thyroid patients. *Clinical Science*, **57**, 375–84.

Wright, P., Gill, M. & Murray, R.M. (1993). Schizophrenia: genetics and the maternal immune response to viral infection. *American Journal of Medical Genetics*, **48**, 40–6.

4 *Biosocial determinants of sex ratios: survivorship, selection and socialization in the early environment*

CAROL WORTHMAN

Introduction

Nature–nurture debates recur in part from underestimation of biosocial dynamics in both evolution and ontogeny. Relationships of sociocultural processes to biological adaptation can be probed by examining afresh an old topic, sex ratio. Close analysis of sex ratio can reveal how social actors and social–cultural construction of environments exert developmental, biological and long-term reproductive fitness consequences. The existence of sexes, and the production and reproductive fitness consequences of sex ratios are central problems for evolutionary biology, while sex ratio is also an object of interest for sociology, public health, reproductive biology, demography, and economics. This chapter argues for a biosocial view of sex ratio determinants, based on analysis of factors that produce sex ratio variation, largely through differential survival. It further promotes a social–ecological view of Darwinian selection by considering fitness consequences of sex-differentiated social treatment and environments.

Sex ratio is essentially an outcome variable, dependent on multiple processes. Most studies of sex ratio variation are couched in terms of long-term outcomes of early environment, namely the fitness consequences (for parent and surviving offspring) of sex-linked manipulation of offspring production and survivorship. Therefore, the present analysis does not seek to consider ultimate causes and proximate mechanisms underlying sex ratio *per se*; rather, it draws on extensive documentation of local variation in, production of, and consequences from, sex ratio variation to illuminate the general biosocial processes that shape human variation in life history. Life history parameters include differential survival and reproduction, developmental schedule, and energetics, but it is argued that intraspecific

44

variation in these is only to be understood in terms of local conditions and decisions shaped through social–historical processes. Thus, consideration of determinants and consequences of sex ratio variation will shed light on human adaptation and the role of life history theory in explaining human variation as an evolved, biosocial product.

Three determinants of sex ratio will be considered: differential production, differential survival, and differential social ecology. Production seems to play a much smaller part in skewing sex ratios than do differential survival and social ecology. Variation in primary and secondary sex ratio in humans is neither systematic nor dramatic; however, postnatal biocultural processes can produce strongly skewed sex ratios of two kinds. One is tertiary sex ratio, a demographic feature that reflects differential care and survival. The other is the sex ratio within which an individual actually operates, and in which socially mediated processes can generate tremendous local variation at adaptively critical points. It is these biocultural processes and their lifetime functional outcomes that will be considered.

Sex ratio

Sex ratio is the number of males relative to the number of females in a population. To be meaningful, the spatiotemporal bases for numerator and denominator must be closely specified. A temporal distinction is made among primary, secondary, and tertiary sex ratio. Primary sex ratio is the sex ratio at conception: in species with internal fertilization and gestation, as with humans, this ratio is necessarily difficult to ascertain directly and must be inferred. Secondary sex ratio is the sex ratio at birth, which results from differential gestational losses compounded with conception ratio. Tertiary, or postnatal, sex ratios arise through differential postnatal losses. The aforementioned definitions are all based on total population numbers of each sex for the condition. But *effective sex ratios*, or the relative number of males and females pertinent to individual fitness, can be highly localized and thus represent a mere fraction of the actual numbers for the entire population. For instance, sex-differentiated local resource competition, such as mate competition, is a function of sex-specific competition for locally available resources in spatially structured populations, and is associated with skewed sex ratios (Hamilton, 1967): individual fitness hinges not on the absolute number of potential mates, but on the number of available, or eligible, mates. The distinction between population sex ratio and effective sex ratio, and determinants of the latter in structured human populations will be important in the discussions below.

Sex allocation

Evolution of sex ratio has concerned biologists since Darwin (1871) first considered the matter. The problem has received close attention for theoretical and practical reasons: sex ratio is a phenotype shared by all sexually reproducing species and directly linked to reproductive strategy, but which is also eminently measurable. Most current work on sex ratio as a phenotypic trait draws upon logic used by Fisher (1930). Fisher pointed out that a 1:1 sex ratio is expectable in diploid dual sex species, because of frequency-dependent natural selection for equal distribution of parental investment which arises from biparental contribution of half the autosomal inheritance to each offspring. Parity of sex ratio follows from parity in sex allocation, on the assumption of equivalent fitness returns by sex per unit of parental investment (Frank, 1987, 1990). Subsequent work has substantially borne out the expected 1:1 ratio (particularly in mammals; Charnov, 1982, p. 115; Clutton-Brock & Albon, 1982), but significant deviations from parity have been observed (Clutton-Brock, 1991, pp. 223–4) and the adaptive reasons for their occurrence have been investigated intensively (Hamilton, 1967; Trivers & Willard, 1973; Charnov, 1982, 1993).

The extensive theoretical and empirical work in this area can be roughly summarized as showing that variation in sex ratio can generally be explained with Fisher's concept modulated by differential reproductive opportunities for males and females (well reviewed in Charnov, 1993). Factors that drive skewed sex ratios include differential parental costs by offspring sex, and conditions creating sex differences in offspring fitness. (Not considered here, but also relevant, are biases in parental contribution through sex-linked inheritance or extra-chromosomal inheritance, largely through cytoplasmic or viral routes, Hamilton (1967); Werren, Nur & Wu, 1988). Sex-differentiated rearing costs can lead to skewed sex ratios simply because, if parents are selected to put equal investment into each sex, they will bear more of the one that is cheaper to rear. Furthermore, if rearing costs differ in kind, parents may skew sex ratio according to their relative ability to meet the needs of either sex. Sex-differentiated offspring fitness arises through characteristics of the parent (ability to rear, status) or of immediate circumstances that exert sex-differentiated effects. Fisher's assumption of panmixia is often violated, with populations showing internal breeding structure and local mate competition (Hamilton, 1967). A non-homogeneous breeding structure can arise directly through barriers (intrinsic or extrinsic, such as physical size or distance) that restrict mate access, or indirectly through differential access to material or social resources that regulate ability to reproduce (e.g. heritable dominance status, class, or territory). In age- , space- , or socially structured breeding populations, population sex ratio reveals little about the adaptive problems and specific responses of constituent subgroups. Consideration of operational sex ratio, defined as the sex ratio of available partners at the time an individual is ready to reproduce, appears to yield better insight into individual behaviour (Clutton-Brock & Parker, 1992).

A more subtle notion of sex ratio emphasizes its contingent and temporal features. The apparent demographical measurability of sex ratio has been something of a conceptual trap, which leads to its treatment solely as a static population trait, a demographical outcome, rather than also as a dynamic, socially regulated and individually differentiated condition. Sex ratio can be viewed as local or individual specific, from which it follows that sex allocation can be understood as a response by reproducers to local conditions, intrinsic or extrinsic to the reproducer, that predict or determine the relative fitness of offspring by sex in a frequency-dependent manner. But ultimate offspring fitness is temporally remote from parental decisions that determine primary or secondary sex ratio, which decisions are thus necessarily probabilistic. Compared to semelparous species which reproduce once, slowly, serially reproducing, or iteroparous, species such as humans may adjust each successive reproductive event to track existing conditions in a way that can shape sex ratio in completed fertility. Skewed sex allocation is more probable when litter size is one, total fertility low, parental care high, and fitness return curves (that is, fitness gains with increasing parental investment) differ by sex (Frank, 1990).

The array of selective forces on sex ratio, surveyed above, has been identified through comparative analysis and is generally assumed to shape intraspecific variation. But this is an empirical question, and consideration of species-specific conditions should inform identification of adaptive processes shaping human sex ratio.

Sex ratio variation in humans

The paucity of effect achieved by tremendous efforts in breeding shows how difficult it is to move mammalian species consistently away from a 1:1 sex ratio (Charnov, 1982; Clutton-Brock & Iason, 1986). Furthermore, the distribution of sibling sex ratio is binomial (Williams, 1979), which does not entirely preclude but strongly mitigates against the ability to adjust sex ratio facultatively according to parental or environmental conditions (Clutton-Brock & Iason, 1986). Secondary sex ratio is relatively easy to ascertain with reasonable accuracy, and is therefore most widely reported. Humans show a systematic, though small, skewing of this ratio, averaging around 105 (males/females × 100) (James, 1987). Male bias has been attributed to the relative cheapness of males in terms of parental investment, since more die early; similarly male-biased secondary sex ratios are frequent among mammals (Clutton-Brock & Albon, 1982). Well-documented cases of secondary sex ratio variation within and between human populations range from around 80 to around 130 (Sieff, 1990); proposed proximate causes of this variation include coital timing and selective gametic or gestational losses (for review see Teitelbaum, 1972; James, 1987; Sieff, 1990). Reports at the extremes of the range are often disputed as products of infanticide rather than true birth ratio, but some well-documented cases of highly skewed ratios exclude infanticide or selective reporting (Hewlett, 1991; Worthman, Stallings & Jenkins, 1993). Variation in the tertiary sex ratio is much more impressive and robust. In a recent survey, Hewlett (1991) found that juvenile (age ≤15 y) sex ratios range from 60 to 172 in demographically documented preindustrial societies; 13 of 47 societies surveyed had juvenile sex ratios over 120, by contrast with the 2 or 16 with birth ratios >120. Small sample size in a number of these populations may increase the contribution of stochastic variation.

But the most dramatic and systematic sex ratio variation among humans has been ignored in this literature. Throughout the life course, human societies structure the effective local sex ratios in which individuals function. In many societies and for much of their lives, people live, work, or socialize in groups with widely skewed sex composition. For instance, it is not unusual for older boys and young men to spend time or live in all-male groups (effective sex ratio = ∞) for initiation, warfare, hunting, recreation, and/or politics. Nor is it unusual for infants and children to be considered as neuter or female by association, and to live in an exclusively maternal or adult female domestic world (sex ratio = 0). In so far as social determinants of sex-, age-, and/or activity-specific sex ratios determine the physical and social environments in which individuals operate, they also determine differential selection (through, for instance, resource availability and use, access to mating opportunity, or exposure to risk). Moreover, an extensive literature documents life-time effects of sex ratios of rearing throughout the juvenile, or prereproductive, period.

In the following sections, two propositions will be considered concerning the contrasting aspects of sex ratio. These are: (i) Population sex ratios, which reflect differential production and survival through selective processes, reflect the degree to which members of a population, by age and sex, represent outcomes of differential selection. They therefore mirror differential environmental quality as well as sex allocation. (ii) Socially determined local sex ratio contributes not only to differential survival but also to differential life history *outcomes*.

Sex differences in humans

Biological sex differences, in humans as in other species, have been explained by natural (including sexual) selection on sex-differentiated roles in reproduction (Frayer & Wolpoff, 1985). Consequently, extant sex differences are thought to provide clues to the differential selection pressures that have operated in the evolution and/or maintenance of these differences. In mammals, obligatory initial investment in each offspring falls largely on females, which has led to the expectation that female fitness will be more constrained by resources, and that of males will be more constrained by females. Sex asymmetries in parental investment drive the competition for mates (the individual who potentially provides more investment is the one competed for), and set the stage for sexual selection driven by preferences of the sought-after sex. Sex differences in humans have been interpreted in this light.

First, men are slightly larger than women by height and weight, although the degree of difference may vary in relation to cultural–ecological conditions (Bogin, 1988; Eveleth & Tanner, 1990, but see Gaulin & Boster, 1985). Body size differences, breast development, and male genital proportions have been attributed to mate competition and sexual selection (Short, 1980; Smith, 1984): while humans have fairly high levels of biparental care (Lancaster & Lancaster, 1987), they also show facultative polygyny (Daly & Wilson, 1983). Variance in reproductive success is greater in males than females.

Secondly, and more important, is dimorphism in body composition, although again cultural–ecological factors may alter the magnitude of sex differences: young American women have 42% greater percentage body fat than young men, and men 23% more percentage muscle than women (Holliday, 1986). In women, the extent and pattern of fat deposition and the sensitivity of reproductive function to indicators of energy supply (indeed, perhaps also reductions in overall size), document the importance of energetics for that sex (Rosetta this volume). Greater muscularity in men may be the product of sex-differentiated productive roles (also related to patterns of parental investment) and sexual selection. Asymmetries in energy constraints arise principally from these differences in size and body composition: across ages and activity states, males generally expend more energy than females per unit body mass (Lowrey, 1978; Schofield, 1985; Spurr & Reina, 1989).

Thirdly, mortality schedules differ in males and females. Under conditions of equivalent care, juvenile males of all ages are more likely to die than females (Daly & Wilson, 1983). The sex difference is greatest in the postnatal period, but persists markedly through weaning to middle childhood. Intrinsic factors, such as developmental immaturity and increased vulnerability to disease (Khoury et al., 1985), are chiefly responsible for greater juvenile male mortality, although accidental death is also greater among boys than girls (World Bank, 1993). Male death through extrinsic factors (mainly war and homicide) rises sharply over the late teens to peak in the mid-twenties (Daly & Wilson, 1983). By contrast, mortality rate in females follows a trajectory parallel to, but lower than that of males in pre-adult years.

Fourthly, females have biologically truncated reproductive lifespans; childbearing ceases on average around age 40, followed by menopause about 8–10 years later (Hill & Hurtado, 1991). Fertility of males declines much less dramatically with age, with decrements from the fifth decade. Thus the temporal window of reproductive opportunity is wider for males than for females, which creates an inherent imbalance in sex ratio of fecund individuals even when numbers and age distribution of men and women are matched (Chagnon, 1988; Clutton-Brock & Parker, 1992).

Lastly, humans exhibit sexual bimaturism, in both the timing and course of puberty (Marshall & Tanner, 1970; Worthman, 1993). Under egalitarian childcare, girls enter puberty 6–12 months ahead of boys, entry being de-fined by onset of sex character development. The slight difference in timing of onset is magnified to a large one with respect to morphological changes of puberty: girls show breast development very soon after ovarian activa-tion, and achieve peak height gains before menarche is attained. Boys, on the other hand, show visible morphological signs of androgenization (muscle development, facial hair, voice change, height gain) after reproduc-tive maturation is well advanced. Current life history theory, which at-tempts to explain variation in life history parameters such as growth rates, reproductive patterns and longevity in evolutionary terms, emphasizes di-rect operation of selection on timing of reproductive maturation to maxi-mize lifetime fitness, under taxon-specific mortality schedules and energy production capacity (Charnov, 1991, 1993; Borgerhoff Mulder, 1992; Hill, 1993). Both mortality schedules and energy production differ by sex in humans and correspond to life history constraints.

Determinants of sex ratios in humans

Sex ratios may be adjusted in various ways. Feasibility, costs and benefits of specific means to effect bias depend on which sex ratio is involved. For instance, mechanisms and attendant cost–benefit balances of manipulating sex ratio at conception differ from those aimed at local mating competition. In this section, sex ratio will be viewed in two ways: as a product of selection, and as local social product. Population sex ratio reflects a series of selection filters created by ecological and social conditions, which operate at different levels and at different times to determine sex- and age-specific mortality. Determinants of six ratio will be examined in this light. Drawing on existing analyses, a hierarchy of possible sex ratio-determining factors in primary, secondary, and tertiary sex ratios is outlined (see also surveys in Otten, 1985; James, 1987; Sieff, 1990).

Determinants of sex ratio at conception

Myriad factors have been implicated, but none definitively demonstrated, as causally linked to primary sex ratio variation. Possible factors include differential survival or motility of sperm (due to properties of sperm themselves and/or biochemistry of the female reproductive tract), and the timing and frequency of coitus (influenced by marriage duration and type, and cohabitation rules) (Guttentag & Secord, 1983, pp. 100–8; James 1986, 1987). A major problem with interpreting the literature on sex ratio is that sex ratio at birth, not ratio at conception, is used for inferring causal relationships, so that differential loss would also count as differential conception. Skewing of sperm production by sex chromosome content is precluded by meiotic production of equal X and Y divisions. Differential survival of sperm by the sex chromosome carried is possible in the male or the female reproductive tract, on the basis of differential weight or distinctive surface properties, such as antigens, on Y-bearing sperm. Biased ratios of sex chromatids in ejaculate have not been demonstrated, but numerous properties of the female reproductive tract may influence sperm survival. These include properties of mucus (pH, viscosity, structure) that influence sperm motility, and the levels of sperm-directed antibodies (James, 1987). Observations of biased sex ratios under specific social conditions suggest a role for coital patterns (Guttentag & Secord, 1983). James (1986) has suggested that sex bias associated with timing of coitus relative to time of ovulation is also due to ratios of gonadotropins to gonadal steroids.

Gestational factors in sex ratio at birth

A systematic review of this topic (Waldron, 1983) found no overall support for the common assumption of systematic male disadvantage *in utero*. Fetal mortality is if anything greater in females through at least the first trimester; greater attrition of males in the last trimester of pregnancy appears to be related to greater male size. Prenatal and obstetric care appear to reduce late fetal male losses.

Sex ratio from birth to onset of reproduction

In this section, general patterns in final causes of mortality are contrasted by sex and age. Social and contextual factors that mediate those patterns are discussed in subsequent sections. Neonatal deaths are around one-third greater among males then females, due largely to greater male immaturity through slower lung maturation and slightly shorter gestational lengths (Waldron, 1983; Johansson & Nygren, 1991). Thereafter, in industrialized societies with an egalitarian ethos, good health care, and life expectancies over 65 years, males experience higher mortality than females throughout the pre-adult years. This discrepancy is attributable to 50% greater accidental and violent death among males than females under 15 years of age in these countries, as well as to increased death from communicable disease (World Bank, 1993, pp. 224–5). Explanations for this pattern have focused on greater male risk-taking behaviour and intrinsic genetic and endocrine factors (Daly & Wilson, 1983; Waldron, 1983); male mortality from injury increases over the second decade and results in widening sex disparities in mortality (Taket, 1986). Relative male vulnerability in immune function, with lower humoral and cell-mediated immunity, has been related both to X-linked immunoregulatory genes with a concomitant Y disadvantage, and to differential endocrine effects on the immune system, wherein, on balance, androgens dampen and estrogens stimulate (surveyed in Otten 1985; Schuurs & Verheul, 1990).

In countries with life expectancies below 60 years, the picture is different and more variable (Waldron, 1987; Murray & Chen, 1994). Infant males experience higher mortality than females in almost all of these countries, except for some with strong son preference (Williamson, 1976), scattered in a band from Northern Africa through Western and Southern Asia. But at ages 1–4 years, female mortality exceeds male mortality in most low life expectancy countries. No single category of mortality risk appears to be responsible; disease (communicable and otherwise) and accidental causes all contribute to varying degrees. Survivorship of girls ages 5–14 years is lower than that of boys only in the band mentioned above; otherwise, excess male mortality is greatest in the late juvenile period, due again to accident and violence. Large worldwide variability in the degree of sex difference in environmental quality and exposure to risks is documented by substantial variation in average countrywide female health advantage (or, conversely, male health disadvantage), from 30% lower disease burden in females of formerly socialist economies, to 8% higher burden in Indian females (World Bank, 1993, p. 28).

Parental care

Sex allocation literature focuses on the distribution of reproductive resources among male and female offspring. In humans, parental care comprises a large component of reproductive effort and crucially determines survival and health of offspring (Daly & Wilson, 1983). This role is dramatically underscored by Ache foragers of Paraguay, among whom maternal death results in 100% mortality if the child is under 2 years of age (against 33% if the mother lives), 36% if under 15 years of age (against 25% if the mother lives); paternal death is associated with 43% mortality for juveniles under 15 years of age, against 19% for those in intact families (Hill & Kaplan, 1988; see also Hurtado & Hill, 1992). Circumstances that constrain parental resources and health can force parental choices in allocation of care, often by sex and/or perceived viability. For instance, Scheper-Hughes (1992) has extensively documented both distal political–economic and proximal infant-specific factors that result in withdrawal of maternal care in a Brazilian *favela*. Dickemann (1979) and Boone (1988) have detailed the effects of social stratification and social position on intrasocietal variation in sex allocation.

The literature on determinants of parental investment in primates in general and humans in particular has been subjected to extensive recent review (Hrdy, 1987, 1990; Sieff, 1990; Clutton-Brock, 1991; Cronk, 1991). This literature suggests a need for attention to the following issues concerning differential care in humans: (1) biased parental care may respond to actual offspring differences in need; (2) differential parental care may aim to shunt children into different available alternate life histories, thereby reducing social niche saturation and/or matching ability to situation; (3) social change or uncertainty can alter differential care patterns according to parental expectations of future opportunities for, and returns from, specific offspring. An example of the third set of conditions prevails among Bundi of Papua New Guinea, where social change increased the value of daughters through introduction of schooling, wage earning, and remittances to home, and resulted in more rapid acceleration in development of girls than boys. This effect probably reflects greater improvements in the quality of the rearing environment for girls than boys (Worthman, 1993; Zemel, Worthman & Jenkins, 1993).

Effects of differential care

Measures to regulate parental investment range across a gradient of severity from infanticide, aggressive neglect, passive or benign neglect, and selective neglect, to differential opportunity or enrichment. Sex ratio only partially reflects this gradient. The extremes of investment represent quite different implications for offspring life history, for at one end (infanticide) there is death or increased risk for death, and at the other end (enrichment) there is resource consolidation and opportunity for enhanced reproductive and/or social success (Hrdy, 1992; Hrdy & Judge, 1993). In-between lies considerable variation in quality of life, in terms both of health and social existence. Individuals and sexes may differ in the vulnerability to decrements in parental care and returns or benefits from increments in care. Despite the prevalent assumption that mortality, morbidity and developmental outcome of males are more sensitive to environmental quality, wide within-population and between-sex variation in environmental quality has precluded a critical test of the notion (Stinson, 1985). 'Environmental quality' is likely an over-general concept that requires closer specification; the sexes differ in *specific* aspects of environmental quality to which they are vulnerable for differential outcomes in survival or development (Worthman, 1993). For example, in the countries with long life expectancy, dramatically reduced mortality from disease in both sexes and accidental causes in girls has resulted in large proportional increases in sex differentials of juvenile mortality due to persistent male risk taking behaviour (Waldron, 1987). Another example is Barker and colleagues' findings of sex-differentiated strength in long-term effects of birthweight on death from cardiovascular disease (Osmond *et al.*, 1993); their studies of long-term effects of early environmental quality on physiology and health bear continued attention with regard to absence and presence of sex differences.

Rearing environments as selection filters

The above reflections on determinants of sex ratio through differential mortality also imply that rearing environments act as selection filters. The higher the mortality at any age, the more strongly selected is the population at later ages. For instance, that the probability of dying before the age of 15 years exceeded 25% in 1950 throughout Africa, Middle East, Asia and the Pacific (World Bank, 1993, p. 203), implies that contemporary adults in these regions are truly survivors. Differential survival may mitigate the apparent long-term impact of differential care on developmental outcomes, because the two sexes have undergone differential selection so that the more selected sex is more refractory to environmental insult. This effect may explain the observation among Hagahai forager–horticulturalists of the New Guinea Highland fringe, that severe early daughter neglect concurrent with a generally harsh rearing environment did not result in differential female delay in endocrine onset of puberty (Worthman *et al.*, 1993). Given the linkage of environmental quality with maturation rates, girls might be expected to delay pubertal onset relative to boys; perhaps girls who survive under these stringent conditions are developmentally canalized.

Gradations in parental care (from infanticide to enrichment) also act as differential selection sieves with differing costs and benefits to parents and their offspring. The more a parent has already invested in a child, the more there is to lose if parental investment decisions impair the child's future. Thus, infanticide is usually an early decision (Daly & Wilson, 1984), but early decisions are costly in terms of probability of error and low specificity with respect to many actual characteristics of a specific child. As the child gets older, abilities and goodness of fit to prevailing and predicted future circumstances can be gauged, and parental resources allocated according-ly. Furthermore, differential allocation of care or opportunity can be based on offspring-intrinsic (for example, sex) or offspring-extrinsic (for instance, birth order) features; only the former will act as a selective pressure, unless the latter interacts with intrinsic factors (such as vulnerability) to determine differential fitness. Overall, parents act as selective agents insofar as differential parental care determines differential offspring fitness.

Effects on life history outcomes

The long-term impact of sex differentiated care on physical development and adult reproductive function has rarely been considered, although its more acute effects on nutritional status and mortality have been heavily documented (Basu, 1989; Muhuri & Preston, 1991). Environmental quality affects both growth rate (height for age) and timing of puberty (Eveleth & Tanner, 1991); sex differences in the timing of puberty can vary across populations (Worthman, 1993), implying variation in sex-differentiated environmental quality. Long-term effects on reproductive function are strongly suggested by comparative data on adolescent development. For instance, among Hagahai, by contrast with western postindustrial socie-ties, puberty commences late (endocrine onset is at approximately 12 years in girls, 13 in boys), is prolonged to a late menstrual onset, and presents low maturational trajectories of reproductive hormones, both regulatory (gonadotropins) and peripheral (ovarian steroids) in females relative to males (Worthman *et al.*, 1993). Lifetime natural fertility of Hagahai women is low (3.8 births; Jenkins, 1987).

This leads to the question of whether, beyond influencing survival, size, and pubertal timing, parents can influence offspring reproductive perform-ance by modulating the quality of the rearing environment. Large population differences in adult levels of ovarian output reflected in luteal progesterone (Ellison *et al.*, 1993) have been tentatively attributed to effects of environment on the ontogeny of endocrine regulation. However, a correlation of population differences in progesterone levels with differences in actual fertility has not been established. Furthermore, endocrine differences between populations or individuals do not necessarily reflect impairment. Rather, such differences may be adaptive because they reflect modulation of endocrine regulatory parameters which enhances phenotypic 'fit' between multiple aspects of reproductive function (such as pregnancy and lactation) and the environments encountered in adulthood, including food supply and workload. Optimal reproductive performance under situations of high female workload, low or uncertain food intake, and high morbidity may be achieved through facultative ontogenetic adjustment of physiologic regulatory parameters which set different functional properties than those developed under conditions of low workload, consistently excellent nutrition, and low morbidity. In so far as the rearing environment influences adult reproductive function, parents may thus affect offspring–environment reproductive fit, or even partially inure surviving young to adverse conditions in adulthood. Again, sexes may differ in their sensitivity to different aspects of adversity.

Whether specific dimensions of parental care have sex-differentiated impact on social viability or behavioural and psychological outcomes has scarcely been investigated. However, it can reasonably be argued that these outcomes are important determinants of adult fitness that produce transgenerational effects on reproductive and parental competence. Reports indicate general sensitivity of parental investment patterns to perceived opportunities and requisites for offspring social or life history success, as manifested in differential care by sex and social status or resources (Dickemann, 1979; Boone, 1988). Within societies, both parental social or resource status and offspring qualities can affect the sexes differently over the life course. Moreover, societies vary in the determinants of male and female social viability and adult fitness. Attention should be given to the titration of parental care with respect to these lifetime, life history effects, beyond simple offspring survivorship or parental provision of reproductive opportunity.

Sociological effects of sex ratio

Guttentag and Secord (1983) have argued that fluctuating population sex ratios have been paralleled by changes in attitudes toward women, sexual mores, marital practices, and family conditions throughout Western history. The authors themselves point out that these shifts occurred within a single general cultural frame including monogamy, agriculture, and social stratification, so the specific effects may not be generalizable.

Nevertheless, the supply of males and females for important social roles, productive and reproductive, and their relative supply to meet role complementarity and other social demands, have important social consequences (Roth, 1985). Cohort or systematic deviations in sex ratio affect the productive base, fertility potential, and mating opportunities, so their effects ramify through life history and across the population.

Effective sex ratios

This essay has so far concerned the conventional sense of sex ratio as a population trait, the ways in which sex ratio variation is generated through biosocial and biocultural processes, and the enduring effects of such processes on individual well-being and life history. Another way to consider sex ratio, is to treat it as a proximal, variable state, that is, in terms of the actual or perceived sex ratio of the social environment for a given individual, or *effective sex ratio*. This concept is similar to, but more general than, the operational sex ratio (Clutton-Brock & Parker, 1992), or the actual pool of available potential mates at the moment an individual is also ready to mate. The effective sex ratio is the sex ratio of pertinent social actors, that which comprises the lived social world of the individual. Societies vary enormously in effective sex ratio in various domains throughout the lifespan, a variation produced through social construction of life history with age- and sex-graded environments in which that history is played out. The range of cultural practices will be considered in this regard, and the notion that systematic, culturally prescribed variations in effective sex ratio exert long-term effects on developmental outcomes will be examined.

The effective sex ratio for infants often approaches zero, which reflects little contact with men, and depends on sex of caregivers and sex composition of domestic and female productive spheres; sex of infant has little influence on these. Fathers are not infrequently absent from the domestic sphere of women and young children: men roomed apart from women and children in 29% of a sample of 159 societies (Whiting & Whiting, 1975; see also Schlegel & Barry, 1991, p. 170). Across societies, from simple to complex, infants have more contact and interaction with girls than boys (Edwards, 1993), most especially in societies with heavy reliance on sibcare. Since infants and young children are often regarded as effectively neuter or part of the 'women's world', their sex is not culturally regarded as contributing to the effective sex ratio, but this ratio exerts sex-differentiated effects on the child's opportunity for same- and opposite-sex interaction. In a cross-cultural study of socialization, Whiting and Edwards (1988, p. 60) found the proportion of adults observed with children ages 5–7 years was female skewed and sex differentiated: the proportion adults present who were female was 69–88% for girls and 49–83% for boys, where figures differ by society.

Effective sex ratio for toddlers and young children through to the age of 5 years is also largely determined by domestic and female productive arrangements. Where mothers leave the home or camp for foraging or field work, alternate caregivers (sibs, kin, other women) are needed, especially if the worksite is dangerous for children or if their presence would markedly impair maternal productivity (Hill & Kaplan, 1988; Hurtado *et al.*, 1992). From the age of 5 years, girls do more responsible productive work than boys, who spend more time in play and unstructured activity (Edwards, 1993). Customary sexual division of labour and living arrangements influence how much young children interact with men over the day. In a six-culture study, proportions of observations of 3–10 year-olds in which fathers were present ranged from 4 to 17%; boys were consistently found with fathers more often than girls in all societies studied (Whiting & Edwards, 1988, p. 59).

Where enough peers are available, sex segregation characterizes middle childhood (ages 6–10 years); from age 5 years, boys aggregate into larger groups of same-sex peers, girls into same-sex dyads and smaller groups (Edwards, 1993; Maccoby, 1988). Concomitantly, boys spend less time with mothers and adult women. Over this period, then, boys experience an increase in effective sex ratio in daily life; for girls it remains low. In some societies, the shift is dramatic: boys are actually moved out of the mother's bedroom or parental house and into a men's or separate boy's house. Such a change may initiate a phase of residence in single-sex groups as part of male initiation cycles (e.g. Herdt, 1982). Removal of boys from the natal unit and participation in all-male residence, work groups, and/or initiation rituals may occur in middle childhood or as late as early-mid adolescence; 27% of traditional societies have men's houses or exclusively male secret societies (Paige & Paige, 1981, p. 232). With the onset of puberty or menses, the social sphere of girls may be more restricted, especially if virginity is valued; of 99 societies sampled, Paige and Paige (1981, p. 235) found menstrual segregation in 31%. Both sexes are usually expected to assume gender-appropriate productive roles by early–mid-adolescence, which involves a shift in effective sex ratio depending on sex composition of work groups. In other words, from late childhood through adolescence, the sexes may experience systematic segregation, so that the effective sex ratio for boys approaches infinity while for girls it approaches zero. Half of the societies sampled by Paige and Paige (1981, p. 238) practise forms of sex segregation, either of women, or men, or both.

Effective sex ratios and life history outcomes

The span of practices in the cultural construction of effective sex ratios by sex across the life course is wide. Sex segregation may be low, and fathers may be intensely involved in domestic life and childcare. Or it may be high, with adolescent and adult years spent in highly sex-segregated social, productive and even reproductive settings (i.e. sex ratio far above 1 for males, below 1 for females). Questions remain concerning whether these arrangements have effects on life history outcomes, and if so, what these effects might be. The following sections will consider possible effects of effective sex ratio on reproduction, production, and psychobehavioural patterns.

Onset of reproductive career

The effective sex ratio in young adulthood is strongly influenced by cultural determinants of onset of reproductive career which, in turn, is often constrained by the productive career, especially in males. This asymmetry leads to the general pattern of later marriage for men than women; the onset of men's reproductive career is less constrained by career duration than is women's. Some societies (17% in one sample; Schlegel & Barry, 1991, p. 39) segregate young men, for short or long intervals, to 'make men of them' or have them 'show their worth' (Paige & Paige, 1981). Whatever the rationalization, the life history effect is to delay the onset of male reproductive career, create a special labour pool, and sift out excess males. Whether in male initiation cycles of New Guinea or in warrior grades of age set, age grade societies of East Africa, young men are effectively removed from the marriage pool for many years, which may provide a demographic basis for polygyny by increasing male marriage age (Chagnon, 1988). Moreover, groups of young men are often allocated high-risk roles, such as raiding, defence, herding, and labour migration, which increase their mortality, expose them to high selection, and remove them from the reproductive pool just before or at onset of reproduction (Dickemann, 1979; Boone, 1988). Finally, performance as a young man often influences social status and resources (Knauft, 1991), which influences fitness in so far as differential social success affects marriageability. The net effect of these practices is to dramatically increase the operational sex ratio for young men, and reduce it for young women and older men. This constrains mate choice of young men and women, and enhances it for older men.

Onset of reproductive career is usually earlier and less uncertain for women than men (extensively surveyed in Whiting, Burbank & Ratner,

1986). Of a 124-society sample, 60% of young women married within 2 years after menarche, while men married on average at least 2 years later (Schlegel & Barry, 1991, p. 40). Effective sex ratio changes less at marriage for females than males, particularly in highly sex-segregated societies, because girls grow up in the female domestic sphere and return to it as wives and mothers, whereas boys may have spent less time in that sphere. Experience with same or opposite sex partners for social interaction appears to influence ease and competence in forming gendered relationships, which suggests that sex ratios experienced in childhood may importantly shape facility and comfort in relating to same and opposite sex persons in adulthood (Maccoby, 1990). Given the prevalence of female exogamy among humans, most young women enter a world of unfamiliar non-kin upon marriage (Rodseth et al., 1991).

Productive ability

The sexual division of labour, a topic of enduring anthropological interest, implies a skewed effective sex ratio of production (D'Andrade, 1966), which is linked to differential risk, energy demands, and possibly, energy intake. Adaptationist explanations for sexual division of labour usually build on constraints inhering to reproductive roles of men and women (Hurtado et al., 1992), and the sex differences in work capacity and metabolism associated with physique and size likely reflect sexual selection for productive performance. Men and their sons may perform intensive agriculture (e.g. Punjab; Mamdani, 1972), or unrelated men in groups may hunt co-operatively (e.g. Ache of Paraguay; Hill & Kaplan, 1988), while women in these groups undertake intensive domestic production with their children or mildly co-operative foraging with unrelated females, respectively. From studies of these and other societies, sex-differentiated labour demands in productive spheres are known to influence reproductive behaviour (including manipulations of sex ratio) and the productive base. But long-term consequences of effective sex ratio in early experience arise directly through learning. In so far as children acquire skills by observation, play, or guided participation with adults (Draper, 1976; Rogoff, 1990), then effective sex ratios shape their opportunities for acquiring knowledge and competence in sex-differentiated arenas. Differential skill acquisition has long-term consequences for life history through its effects on productive capacities. Differential productive tasks of children provide not only opportunities for learning, but also exposure to risk: for instance, sex-specific task allocation (not just differential care) may explain wide variation among countries in accident rates among girls under 15 years of age (Waldron, 1987).

Psychosocial development

A substantial literature in anthropology and developmental psychology has considered the effects of the effective sex ratio of rearing, and especially of parental roles in socialization, on acquisition of social skills and attitudes toward self and others (Hoffman, 1988; Hewlett, 1992; Maccoby, 1992; Darling & Steinberg, 1993). Drawing on Freudian concepts, anthropological research has focused especially on effects of father absence (Whiting & Whiting, 1975; Munroe, Munroe & Whiting, 1981; Herdt, 1989). J. Whiting proposed that sex identity conflict occurs in societies with low male salience in infancy, where boys reared in a maternal-dominant domestic sphere identify first with women perceived as powerful, and then experience identity conflict when they realize that men actually hold sociopolitical power (for review see Munroe *et al.*, 1981). Sex identity conflict was expected to affect men's sexuality and attitude toward women, and to be expressed through wide-ranging effects on customary practice. Because of limitations of cross-cultural comparative method and non-specificity in the concept of sex identity conflict, a causal link between father absence and such conflict has been difficult to establish (Broude, 1981; Herdt, 1989). Herdt (1989), by contrast, has emphasized the need for attention to the dynamics as well as ecology of early experience, and suggested that ambivalent proximity to father (physically close but socially distant) better explains the case of ritualized homosexuality in Melanesia. While long-term effects of early experience on reproductive behaviour and life history are receiving fresh theoretical treatment from an adaptationist perspective (Draper & Harpending, 1982; Belsky, Steinberg & Draper, 1991; Chisholm, 1993), the relative importance of early versus later, concurrent experience remain hotly debated. Indeed, the role of the family in socialization of personality and behaviour is currently in dispute by virtue of recognition of substantial sibling variation in outcomes (Plomin & Daniels, 1987).

Effects of 'the company they keep' (Edwards, 1993) on children's acquisition of cognitive and social skills have been more clearly demonstrated (Whiting & Edwards, 1988), but the linkage to differential life history outcomes has not been investigated. Although much socialization practice and skewing of effective sex ratio is frankly geared to production of differential knowledge, skills (productive and social), and social bonds, remarkably little is known about outcomes in terms of differential survival, fitness, or other aspects of life history. Males appear to seek and obtain greater opportunity to forge bonds with other males, which are not disrupted by marriage and promote co-operative social relations in fraternal interest groups in adulthood (Paige & Paige, 1981; Rodseth *et al.*,

1991; Smuts, 1992). Females, on the other hand, spend less time with peers and usually experience disruption of social bonds through exogamous marriage (Rodseth *et al.*, 1991).

Conclusions

In this chapter, the long-term effects of sex ratio have been considered in lifespan and life history terms. Although secondary sex ratio in humans has not been shown to deviate systematically from around 105, postnatal sex ratios vary widely and reflect differential mortality rates that are influenced by biocultural determinants of sex- and age-specific environmental quality. Skewed tertiary sex ratios reflect differential survival, and thus differential selection, so that sex-differentiated care can be viewed as setting up differential selection screens through which either sex passes. Since some key biological parameters may differ by sex, it is the interaction of sex-differentiated biology with sex-differentiated environment under different fitness constraints that generate both asymmetric selection and asymmetric developmental outcomes. Furthermore, the degree to which members of either sex experience different environmental quality and morbidity/mortality schedules may result in sex-differentiated modulation of life history parameters. Such differential life history modulation is indicated by population variation in degree of sex difference in timing of puberty (Worthman, 1993). Sex-differentiated effects on life history in terms of reproductive career, adult morbidity, and aging have scarcely been explored, although the influence of gestational and postnatal environmental quality has been linked to the life history parameter of maintenance, reflected in long-term effects on metabolism, immune function and organic (cardiovascular and lung) disease (Osmond *et al.*, 1993; see Barker, this volume). Additionally, population sex ratios affect the relative availability of each sex to fulfil productive and reproductive roles, and thus have individual life history and population level effects.

This analysis has sought to go beyond the usual treatment of sex ratio by considering the ecologically and phenomenologically salient, or effective, sex ratio of rearing in terms of life history outcomes. Human societies clearly and systematically skew the sex ratio of specific settings in which children grow up, and often in a sex-differentiated manner. The two areas in which effective sex ratios most clearly exert long-term life history effects are in skill acquisition and mating opportunity or onset of reproduction. The sex ratio of settings in which children participate can, in as much as there is a sexual division of labour and social life, determine opportunities to observe and practice specific productive and social skills, as well as to form long-term relationships. Skill acquisition is crucial for productive capacity and social viability, both of which influence fitness. Moreover, marriage rules, sex segregation and a male youth stage can contribute to highly skewed operational sex ratios for young men and reduce them for young women and older men, with implications for reproductive life history.

Biologists have focused on too narrow a range of sex ratio phenomena, and thus on too narrow a range of sex-differentiated selective pressures. Human societies everywhere view and treat members of each sex differently, although the magnitude and type of differential treatment varies widely. Consideration of processes determining tertiary and effective sex ratios have shown social actors as active selective agents. Sex-differentiated care, task, role and status allocation, and overall social construction of gender-differentiated social and ecological niches set up sex-differentiated selective pressures and adaptative responses which may be termed *gender selection*. Gender selection differs from sexual selection in that it involves not mate competition or choice, but differential treatment by other social actors based on perceived gender. The concept of gender selection may generalize to any social species. This term may be useful in so far as it redresses general inattention to the selection potential of differential treatment by conspecifics, which differential treatment affects the type and strength of selection pressures to which the sexes are subject across the course of life.

The patterns discussed here are suggestive, but more empirical support would be required to ascertain whether effective sex ratios are useful predictors of life history outcomes, and whether gender selection will help explain sex differences. Nevertheless, both terms point to the need for more individual-centered, less population-based microecological concepts for understanding systematic adaptive bases for human variation. Furthermore, they indicate the potential importance of social actors who may treat members of either sex very differently and thereby act as selective forces by strongly shaping both survival and phenotypic outcome.

Acknowledgements

I thank my colleague, Mark Ridley, for critical and bibliographic suggestions.

References

Basu, A.M. (1989). Is discrimination in food really necessary for explaining sex differentials in childhood mortality? *Population Studies*, **43**, 193–210.
Belsky, J., Steinberg, L. & Draper, P. (1991). Childhood experience, interpersonal

development, and reproductive strategy; an evolutionary theory of socialization. *Child Development*, **62**, 647–70.

Bogin, B. (1988). *Patterns of Human Growth*. Cambridge: Cambridge University
Press.

Boone, J. (1988). Parental investment, social subordination, and population
processes among the 15th and 16th Century Portugese nobility. In *Human
Reproductive Behavior*, ed. L. Betzig, M. Borgerhoff-Mulder and P. Turke,
pp. 23–48. Cambridge: Cambridge University Press.

Bergerhoff Mulder, M. (1992). Reproductive decisions. In *Evolutionary Ecology
and Human Behavior*, ed. E.A. Smith and B. Winterhalder, pp. 339–74. New
York: Aldine de Gruyter.

Broude, G. (1981). The cultural management of sexuality. In *Handbook of
Cross-cultural Human Development*, ed. R.H. Munroe, R.L. Munroe and B.B.
Whiting, pp. 633–73. New York: Garland Press.

Chagnon, N.A. (1988). Male Yanomamö manipulations of kinship classifications
of female kin for reproductive advantage. In *Human Reproductive Behavior*,
ed. L. Betzig, M. Borgerhoff-Mulder and P. Turke, pp. 23–48. Cambridge:
Cambridge University Press.

Charnov, E.L. (1982). *The Theory of Sex Allocation*. Princeton: Princeton
University Press.

Charnov, E.L. (1991). Evolution of life history variation among female mammals.
Proceedings of the National Academy of Sciences, **88**, 1134–7.

Charnov, E.L. (1993). *Life History Invariants*. Oxford: Oxford University Press.

Chisholm, J.S. (1993). Death, hope, and sex: life-history theory and the development of reproductive strategies. *Current Anthropology*, **34**, 1–24.

Clutton-Brock, T.H. (1991). *The Evolution of Parental Care*. Princeton: Princeton
University Press.

Clutton-Brock, T.H. & Albon, S.D. (1982). Parental investment in male and female
offspring in mammals. *Current Problems in Sociobiology*, pp. 223–48. Cambridge: Cambridge University Press.

Clutton-Brock, T.H. & Iason, G.R. (1986). Sex ratio variation in mammals.
Quarterly Review of Biology, **61**, 339–74.

Clutton-Brock, T.H. & Parker, G.A. (1992). Potential reproductive rates and the
operation of sexual selection. *Quarterly Review of Biology*, **67**, 437–56.

Cronk, L. (1991). Preferential parental investment in daughters over sons. *Human
Nature*, **2**, 387–417.

Daly, M. & Wilson, M. (1983). *Sex, Evolution and Behavior*, 2nd edn. Boston: PWS
Publishers.

Daly, M. & Wilson, M. (1984). A sociobiological analysis of human infanticide. In
Infanticide, ed. G. Hausfater and S.B. Hrdy, pp. 439–62. New York: Aldine.

D'Andrade, R. (1966). Sex differences and cultural institutions. In *The Development
of Sex Differences*, ed. E. Maccoby, pp. 174–204. Stanford: Stanford University Press.

Darling, N. & Steinberg, L. (1993). Parenting style as context: an integrative model.
Psychology Bulletin, **113**, 487–96.

Darwin, C. (1871/1981). *The Descent of Man, and Selection in Relation to Sex*.
Princeton: Princeton University Press.

Dickemann, M. (1979). Female infanticide, reproductive strategies, and social
stratification: a preliminary model. In *Evolutionary Biology and Human Social*

Behavior, ed. N.A. Chagnon and W. Irons, pp. 321–67. North Scituate, MA: Duxbury Press.

Draper, P. (1976). Social and economic constraints on child life among the !Kung. In *Kalahari Hunter-Gatherers*, ed. R.B. Lee and I. DeVore, pp. 199–217.

Draper, P. & Harpending, H. (1982). Father absence and reproductive strategy: an evolutionary perspective. *Journal of Anthropological Research*, **38**, 255–73.

Edwards, C.P. (1993). Behavioral sex differences in children of diverse cultures: the case of nurturance to infants. In *Juvenile Primates: Life History, Development and Behavior*, ed. M. Pereira & L. Fairbanks, pp. 327–38. Oxford: Oxford University Press.

Ellison, P.T., Panter-Brick, C., Lipson, S.F. & O'Rourke, M.T. (1993). The ecological context of human ovarian function. *Human Reproduction*, **8**, 2248–58.

Eveleth, P.B. & Tanner, J.M. (1990). *Worldwide Variation in Human Growth*. Cambridge: Cambridge University Press.

Fisher, R.A. (1930). *The Genetical Theory of Natural Selection*. Oxford: Clarendon Press.

Frank, S.A. (1987). Individual and population sex allocation patterns. *Theoretical Population Biology*, **31**, 47–74.

Frank, S.A. (1990). Sex allocation theory for birds and mammals. *Annual Review of Ecology and Systematics*, **21**, 13–55.

Frayer, D. & Wolpoff, M.H. (1985). Sexual dimorphism. *Annual Review of Anthropology*, **14**, 429–73.

Gaulin, S. & Boster, J. (1985). Cross-cultural differences in sexual dimorphism: is there any variance to be explained? *Ethology and Sociobiology*, **6**, 219–25.

Guttentag, M. & Secord, P.F. (1983). *Too Many Women?* Beverly Hills: Sage.

Hamilton, W.D. (1967). Extraordinary sex ratios. *Science*, **156**, 477–88.

Herdt, G.H., ed. (1982). *Rituals of Manhood*. Berkeley: University of California Press.

Herdt, G.H. (1980). Father presence and ritual homosexuality: paternal deprivation and masculine development in Melanesia reconsidered. *Ethos*, **17**, 326–70.

Hewlett, B. (1991). Demography and childcare in preindustrial societies. *Journal of Anthropological Research*, **47**, 1–37.

Hewlett, B., ed. (1992). *Father–Child Relations*. New York: Aldine de Gruyter.

Hill, K. (1993). Life history theory and evolutionary anthropology. *Evolutionary Anthropology*, **2**, 78–88.

Hill, K. & Hurtado, A.M. (1991). The evolution of premature reproductive senescence and menopause in human females. *Human Nature*, **2**, 313–50.

Hill, K. & Kaplan, H. (1988). Tradeoffs in male and female reproductive strategies among the Ache. Parts I & II. In *Human Reproductive Behavior*, ed. L. Betzig, M. Borgerhoff Mulder and P. Turke, pp. 277–89, 291–305. Cambridge: Cambridge University Press.

Hoffman, L.W. (1988). Cross-cultural differences in child-rearing goals. In *Parental Behavior in Diverse Societies*, ed. R.A. LeVine, P.M. Miller and M.M. West, pp. 99–122. San Francisco: Jossey-Bass.

Holliday, M.A. (1986). Body composition and energy needs during growth. In *Human Growth*, 2nd ed., Vol. 2, ed. F. Falkner and J.M. Tanner, pp. 101–45. New York: Plenum.

Hrdy, S.B. (1987). Sex-biased parental investment among primates and other mammals: a critical evaluation of the Trivers–Willard hypothesis. In *Child Abuse and Neglect*, ed. R.J. Gelles and J.B. Lancaster, pp. 97–147. New York: Aldine de Gruyter.

Hrdy, S.B. (1990). Sex bias in nature and in history; a late 1980s reexamination of the 'biological origins' argument. *Yearbook of Physical Anthropology*, **33**, 25–37.

Hrdy, S.B. (1992). Fitness tradeoffs in the history and evolution of delegated mothering with special reference to wet-nursing, abandonment, and infanticide. *Ethology and Sociobiology*, **13**, 409–42.

Hrdy, S.B. & Judge, D.S. (1993). Darwin and the puzzle of primogeniture. *Human Nature*, **4**, 1–45.

Hurtado, A.M. & Hill, K.R. (1992). Paternal effect on offspring survivorship among Ache and Hiwi hunter-gatherers: implications for modeling pair-bond stability. In *Father–Child Relations*, ed. B. Hewlett, pp. 31–76. Hawthorne, NY: Aldine de Gruyter.

Hurtado, A.M., Hill, K., Kaplan, H. & Hurtado, I. (1992). Trade-offs between female food acquisition and child care among Hiwi and Ache foragers. *Human Nature*, **3**, 185–216.

James, W. (1986). Hormonal control of the sex ratio. *Journal of Theoretical Biology*, **118**, 427–41.

James, W. (1987). The human sex ratio, 1: A review of the literature. *Human Biology*, **59**, 721–52.

Jenkins, C.L. (1987). Medical anthropology in the Western Schrader Range, Papua New Guinea. *National Geographic Research*, **3**, 412–30.

Johansson, S. & Nygren, O. (1991). The missing girls of China: a new demographic account. *Population and Development Review*, **17**, 35–51.

Khoury, M.J., Marks, J.S., McCarthy, B.J. & Zaro, S.M. (1985). Factors affecting the sex differential in neonatal mortality: the role of respiratory distress syndrome. *American Journal of Obstetrics and Gynecology*, **151**, 777–82.

Knauft, B.M. (1991). Violence and sociality in human evolution. *Current Anthropology*, **32**, 391–428.

Lancaster, J.B. & Lancaster, C.S. (1987). The watershed: change in parental-investment and family-formation strategies in the course of human evolution. In *Parenting Across the Life Span*, ed. J.B. Lancaster, A. Rossi, J. Altmann and L. Sherrod, pp. 187–205. New York: Aldine de Gruyter.

Lowrey, G.H. (1978). *Growth and Development of Children*, 7th ed. Chicago: Year Book Medical.

Maccoby, E.E. (1988). Gender as a social category. *Developmental Psychology*, **24**, 755–65.

Maccoby, E.E. (1990). Gender and relationships. *American Psychologist*, **45**, 513–20.

Maccoby, E.E. (1992). The role of parents in the socialization of children: an historical overview. *Developmental Psychology*, **28**, 1006–17.

Mamdani, M. (1972). *The Myth of Population Control*. New York: Monthly Review.

Marshall, W.A. & Tanner, J.M. (1970). Variations in the pattern of pubertal changes in boys. *Archives of Disease in Childhood*, **45**, 13–23.

Muhuri, P.K. & Preston, S.H. (1991). Effects of family composition on mortality differentials by sex among children in Batlab, Bangladesh. *Population and*

Development Review, **17**, 415–34.

Munroe, R.L., Munroe, R.H. & Whiting, J.W.M. (1981). Male sex-role resolutions. In *Handbook of Cross-Cultural Human Development*, ed. R.H. Munroe, R.L. Munroe and B.B. Whiting, pp. 611–32. New York: Garland Press.

Murray, C.J.L. & Chen, L.C. (1994). Dynamics and patterns of mortality change. In *Health and Social Change in International Perspective*, ed. L.C. Chen, A. Kleinman and N.C. Ware, pp. 3–23. Boston: Harvard School of Public Health.

Osmond, C., Barker, D.J.P., Winter, P.D., Fall, C.H.D. & Simmonds, S.J. (1993). Early growth and death from cardiovascular disease in women. *British Medical Journal*, **307**, 1519–23.

Otten, C.M. (1985). Genetic effects on male and female development and on the sex ratio. In *Male-Female Differences: A Bio-Cultural Perspective*, ed. R.L. Hall, pp. 155–217. New York: Praeger.

Paige, K. & Paige, R.L. (1981). *Politics of Reproductive Ritual.* Berkeley: University of California Press.

Plomin, R. & Daniels, D. (1987). Why are children in the same family so different from one another? *Behavioral and Brain Sciences*, **10**, 1–60.

Rasa, A., Vogel, C. & Voland, E., eds. (1989). *The Sociobiology of Sexual and Reproductive Strategies.* London: Chapman and Hall.

Rodseth, L., Wrangham, R.W., Harrigan, A.M. & Smuts, B.B. (1991). The human community as a primate society. *Current Anthropology*, **32**, 221–54.

Rogoff, B. (1990). *Apprenticeship in Thinking.* New York: Oxford University Press.

Roth, E.A. (1985). Population structure and sex differences. In *Male–Female Differences: A Bio-Cultural Perspective*, ed. R.L. Hall, pp. 219–98. New York: Praeger.

Scheper-Hughes, M. (1992). *Death without Weeping.* Berkeley: University of California Press.

Schlegel, A. & Barry, H. (1991). *Adolescence: An Anthropological Inquiry.* New York: Free Press.

Schofield, W.M. (1985). Predicting basal metabolic rate, new standards and review of previous work. *Human Nutrition Clinical Nutrition*, **39C**, Suppl. 1, 5–41.

Schuurs, A.H.W.M. & Verheul, H.A.M. (1990). Effects of gender and sex steroids on the immune response. *Journal of Steroid Biochemistry*, **35**, 157–72.

Short, R. (1980). The origins of human sexuality. In *Human Sexuality*, ed. C.R. Austin and R.V. Short, pp. 1–33. *Reproduction in Mammals*, vol. 8. Cambridge: Cambridge University Press.

Sieff, D.F. (1990). Explaining biased sex ratios in human populations. *Current Anthropology*, **31**, 25–48.

Smith, R.L. (1984). Human sperm competition. In *Sperm Competition and the Evolution of Animal Mating Systems*, ed. R.L. Smith, pp. 601–59. Orlando: Academic Press.

Smuts, B. (1992). Male aggression against women: an evolutionary perspective. *Human Nature*, **3**, 1–44.

Spurr, G.B. & Reina, J.C. (1989). Energy expenditure/basal metabolic rate ratios in normal and marginally undernourished Colombian children 6–16 years of age. *European Journal of Clinical Nutrition*, **43**, 515–27.

Stinson, S. (1985). Sex differences in environmental sensitivity during growth and development. *Yearbook of Physical Anthropology*, **28**, 123–47.

Taket, A. (1986). Accident mortality in children, adolescents and young adults.

World Health Statistics Quarterly, **40**, 232–56.

Teitelbaum, T. (1972). Factors associated with the sex ratio in human populations. In *The Structure of Human Populations*, ed. G.A. Harrison and A.J. Boyce, pp. 90–109. Oxford: Clarendon Press.

Trivers, R.L. & Willard, D.E. (1973). Natural selection of parental ability to vary the sex ratio of offspring. *Science*, **179**, 90–2.

Waldron, I. (1983). Sex differences in human mortality: the role of genetic factors. *Social Science and Medicine*, **17**, 321–33.

Waldron, I. (1987). Patterns and causes of excess female mortality among children in developing countries. *World Health Statistics Quarterly*, **40**, 194–210.

Werren, J.H., Nur, U. & Wu, C-I. (1988). Selfish genetic elements. *Trends in Ecology and Evolution*, **3**, 297–302.

Whiting, B.B. & Edwards, C.P. (1988). *Children of Different Worlds*. Cambridge: Harvard University Press.

Whiting, J.W.M. & Whiting, B.B. (1975). Aloofness and intimacy of husbands and wives. *Ethos*, **3**, 183–207.

Whiting, J.W.M., Burbank, V.K. & Ratner, M.S. (1986). The duration of maidenhood across cultures. In *School-Age Pregnancy and Parenthood*, ed. J.B. Lancaster and B.A. Hamburg, pp. 273–302. New York: Aldine de Gruyter.

Williams, G.C. (1979). The question of adaptive sex ratio in outcrossed vertebrates. *Proceedings of the Royal Society of London, series B*, **205**, 567–80.

Williamson, N.E. (1976). *Sons or Daughters*. Beverly Hills: Sage.

World Bank (1993). *World Development Report*. New York: Oxford.

Worthman, C.M. (1993). Bio-cultural interactions in human development. In *Juvenile Primates: Life History, Development and Behavior*, ed. M. Pereira and L. Fairbanks, pp. 339–58. Oxford: Oxford University Press.

Worthman, C.M., Stallings, J.F. & Jenkins, C.L. (1993). Developmental effects of sex-differentiated parental care among Hagahai foragers. *American Journal of Physical Anthropology Suppl.*, **16**, 212.

Zemel, B., Worthman, C.M. & Jenkins, C. (1993). Differences in endocrine status associated with urban–rural patterns of growth and maturation in Bundi (Gende-speaking) adolescents of Papua New Guinea. In *Urban Ecology and Health in the Third World*, ed. L.M. Schell, M.T. Smith and A. Bilsborough, pp. 38–60. Cambridge: Cambridge University Press.

5 *Antenatal and birth factors and their relationships to child growth*

NOËL CAMERON

Introduction

The acquisition of evidence for *long-term* effects of the antenatal environment and birth factors on human growth is fraught with difficulty. If the question of *long-term consequences* and *phenotypic expression* is to be addressed, it is necessary to look beyond the perinatal period and attempt to correlate the effects of the maternal environment prior to, and at, birth with growth and physical status at some time following infancy. Such an investigation is by no means easy. Ideally, it requires longitudinal data collected over a considerable timespan (depending on the practical definition of 'long-term') on a sample of known genotype. In addition, it requires the researcher to monitor carefully and account for (control statistically) pre- and postnatal environmental factors. Furthermore, the presence of a genotype–environment interaction, within the complete expression of phenotypic variance, cannot be ignored because the variance of this interaction is an important, but rarely accounted for, component of total phenotypic variance.

Such studies on humans are rare, not only because of the obvious logistical problems of maintaining a longitudinal study with a statistically viable sample size, but also because any such sample is characteristically going to be highly selected and thus environmentally homogenous. That homogeneity works to inflate associations falsely between related pairs of individuals, making the interpretation of a statistical association difficult (Suzanne, 1980; Mueller, 1987). Animal models, on the other hand, have the potential for control of both the genotype and the environment. Thus by holding genotype constant, through selective breeding programmes, and manipulating the maternal environment, the consequences of environmental change on the phenotype of the offspring can be investigated. Such studies have provided unequivocal evidence that the maternal environment during pregnancy and lactation affects not only the initial rate of

69

development of the young, but also the final adult size of the offspring (Dubos, Savage & Schaedler, 1966).

Carefully designed and controlled large-scale birth-cohort studies on humans have overcome some of these design problems and produced notable results. In particular, the 1954 Kauai birth-cohort study of Werner & Smith (1982) in which an island population in Hawaai was studied for almost two decades, and the Dunedin birth-cohort study (Silva, 1990) initiated in 1972, have followed significant samples for over 20 years.

A second strategy is to use 'follow-up' studies that involve the long-term follow-up of the subjects of maternal and child studies originally carried out three or four decades ago. Of particular importance are recent reports that involve investigations on the offspring of adults who were subjects in a variety of maternal and child health studies carried out in the 1950s and 1960s. Alberman et al. (1992), for instance, have recently published on the subjects of the British National Birthday Trust Fund cohort of 1958. Similarly, but of shorter duration, are follow-up studies on the American Collaborative Perinatal Study (Heikkinen et al., 1992), and the Project on Preterm and Small for gestational age infants (POPS) of the Netherlands (Schreuder et al., 1992), in addition to numerous studies that trace individuals from hospital and national birth registration records (e.g. Law et al., 1992).

The alternative to longitudinal birth-cohort and follow-up studies is to approach the problem by using a cross-sectional design. The major disadvantage of such a design is to find strategies to control for different environmental effects on the different age cohorts. However, cross-sectional designs have the advantages of a large sample size and environmental heterogeneity and thus greater variability with which to test different hypotheses.

These approaches have provided considerable evidence that some aspects of the antenatal environment persist into adult life, and may even have a persistent effect in the next generation. Lumey (1992), for instance, investigated the birthweights of Dutch infants born to mothers who were themselves exposed to famine in utero during the 'hunger winter' of 1944–1945. Women exposed to famine during their first and second trimester had offspring with birthweights lower than mothers not exposed to famine, but the birthweights of offspring from those exposed to famine during their third trimester were not reduced. Leff et al. (1992) demonstrated that it is 80% more likely for a low birthweight (LBW) child, compared to a normal birthweight child, to have a LBW mother. In other words, clear effects on reproductive outcome are seen in the generation following an environmental exposure in utero. Werner & Smith (1982) reported that survivors of severe perinatal stress in the Kauai birth cohort study had persistent physical, learning, or mental health problems at 18 years of age. The rate of mental retardation in this group was ten times, the rate of serious mental health problems was five times, and the rate of serious physical handicaps was more than twice, that found in the total cohort. Among survivors of moderate perinatal stress, the rates of mental health problems and mental retardation were two to three times that of the total cohort, but physical handicaps rates were the same.

Even when the interuterine environment is not stressful, intergenerational effects have been reported. Alberman *et al.* (1992), for instance, have demonstrated a significant intergenerational effect on birthweight of children whose parents were subjects in the 1958 British National Birthday Trust Fund cohort (BNBTF). Long-term effects of the antenatal environment on adult physique, and disease risk, are also beginning to emerge. Again, using data from the BNBTF, Alberman *et al.* (1991) have demonstrated that birthweight and maternal prepregnancy weight contribute significantly towards an intergenerational positive secular trend in adult height. Barker *et al.* (1989) have related interuterine growth retardation to the risk of adult hypertension. In a follow-up study of men born between 1920 and 1943 in the United Kingdom, Law *et al.* (1992) demonstrated that the tendency to store fat abdominally, which is known to increase the risk of cardiovascular disease and diabetes, independently of obesity, is related to birthweight and may be a persisting response to adverse conditions and growth failure in fetal life and infancy.

Recognition of *normal* growth and adult physical characteristics is based on the results from a variety of growth studies of children living in developed countries. Identification of abnormal or unusual growth patterns is thus related to the extremes of normality to be found in such countries. Children born within the normal distribution of birthweights in these countries, who subsequently enjoy a reasonably stress-free environment, demonstrate a growth pattern that is used as a reference for other less developed countries. The fact that secular trends for increasing birthweight (Cameron, 1991) and greater adult statures are reported frequently from developing countries and, indeed, that the presence of a positive secular trend and the rate at which it changes, is used as a monitor of improving social and health conditions, implies that the growth of children in developing countries can be used as a gauge of the effects on growth of a poor antenatal environment in addition to chronic environmental stress throughout childhood.

Pregnant women in developed countries enjoy the protection offered by a variety of support services that mitigates against the antenatal environment adversely affecting the fetus. Such protection is minimal or absent for women in the developing world. These women are likely to be from an ethnic group other than European. Even in developed countries, non-European women have a distinctly greater risk of delivering a child of low birthweight (Eisner *et al.*, 1979; Kleinman & Kessel, 1987; Shiono *et al.*, 1986). Women in developing countries are likely to live in a rural area, with little or no access to electricity, clean water, and/or sanitation. Whilst their diet may change during pregnancy, it is unlikely that they will receive supplementation to make up for the energy drain caused by the growing fetus. They are unlikely to be educated, by 'western standards', beyond a base level and they are unlikely to have access to anything other than primary health care. Indeed, once pregnant it is more likely that they will be subjected to a variety of 'traditional practices' with regard to their health and hygiene during pregnancy that may, or may not, be beneficial to the health of their children. They will probably not receive regular antenatal care and are most likely, if their pregnancy appears normal, to give birth at home with the help of a 'traditional' birth attender. It is unlikely that such mothers will smoke, or drink alcohol, or have access to drugs such as caffeine, cocaine, and heroin that have been the focus of attention of many studies in the developed world. The 'husbands' of such women are also likely to be uneducated beyond a base level, although they have a better chance of being educated than their wives. If they are employed, these men may be agricultural workers or migrant labourers, and thus their families will be subjected to a seasonal variation in financial support and food availability. In addition, women and children may not be protected from the annual cycle of disease associated with seasonal climatic change.

Postnatal growth

The growth pattern that is usually described as 'normal' is based on the growth of children who live in developed countries and whose mothers, therefore, were not subjected to many of the adverse antenatal factors, described above, and which are inherent in developing countries. The pattern of 'normal' growth exhibited by children from developed countries prior to birth is characterized by a peak in length velocity between 20 and 30 post-last menstrual period (LMP) weeks and a peak in weight velocity between 30 and 40 post-LMP weeks. Normal birthweight, based on the 1980–1987 United States Birth Records (CDC Atlanta unpublished report), is 3570 g for males and 3410 g for females. Following birth, growth velocity in both height and weight gradually diminishes until a peak is observed between six and eight years in both sexes, known as the mid-growth or juvenile growth spurt. A second growth spurt, the adolescent growth spurt, occurs during the teenage years, but demonstrates sexual

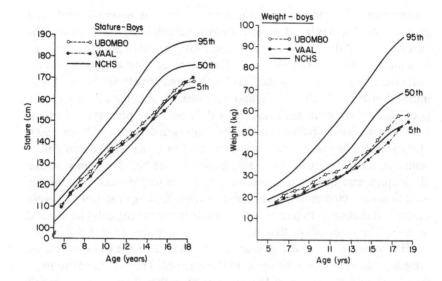

Figure 5.1. Mean heights and weights of boys from Ubombo and Vaalwater, South Africa relative to the NCHS reference charts.

dimorphism in that females exhibit the spurt on average two years prior to males.

The effect on the growth of children of an adverse antenatal environment, combined with continued environmental stress throughout childhood, may be illustrated by a comparison of the growth of children from developed and developing countries. This pattern of growth, in both distance and velocity, is known to differ in magnitude and timing depending on the genetic and environmental characteristics of the source sample (Cameron, 1991). A comparison of two samples from dramatically different environments, and different gene pools, will serve as an extreme illustration.

Figure 5.1 compares the growth of rural South African black children with the American National Center for Health Statistics (NCHS) references. In terms of somatic growth, comparisons are traditionally carried out using a set of 'standard' data as the 'norm' or 'ideal' growth to be expected in an optimal environment. The NCHS charts (Hamill *et al.*, 1979) are the usual reference charts for international comparison, but many countries in the developed world, and some in the developing world, have their own *national* reference charts to enable a more sensitive comparison (Cameron, 1986; Cole, 1993). The advantage of the latter charts is the assumption of a common environment, at least at national level, and greater local knowledge of potential environmental effects. However, the vast majority of the available information with regard to growth in infancy is related to the NCHS references. The reference data relating to infant years were largely based on the Fels longitudinal growth study which was initiated in 1929 (Roche, 1992), and have recently become the subject of a good deal of criticism with regard to their accurate portrayal of infant growth in modern populations (Wright, Waterson & Aynsley-Green, 1993; Dettwyler & Fishman, 1992). Criticisms by Dettwyler & Fishman (1992) centre on the fact that the Fels sample used for the NCHS references, like all samples, was specific to a certain time and place. Mothers and infants were thus subjected to antenatal care and child-feeding practices that were deemed, at that time, to be most appropriate in promoting child health and growth. The recognition that many modern samples do not follow the NCHS norms in the first few years of life has raised the question of the accuracy and thus appropriateness of these norms. The NCHS norms must thus be viewed as *references* for comparison rather than as ideal growth *standards* that judge the *normality* of growth.

The South African rural children in this example live in Ubombo, Kwazulu, in the north-eastern region of South Africa and in Vaalwater (Vaal) in the north-west of South Africa, and have been the subjects of longitudinal growth studies carried out between 1985 and 1991 (Cameron, 1991). The Kwazulu environment is characterized by a major seasonal fluctuation in food and money supply; they live in simple mud and brick houses and have little or no access to electricity and running water. By 6 years of age their mean heights and weights approach the NCHS tenth centile and continue to decline such that 50% of the sample are below the fifth centile by the initiation of adolescence. The adolescent growth spurt is extended on the time axis such that the children exhibit a lengthened, but not necessarily delayed, growth spurt and continue growing into their twenties to end up with mean heights and weights that fall between the NCHS 25th and 50th centiles. Such a growth pattern is characteristic of children in developing countries and it has been suggested (Cameron, Gordon-Larsen & Wrchota, 1994) that this adolescence growth pattern 'compensates' for chronic environmental effects on preadolescent growth. Studies of secondary sexual development demonstrate that, in rural children of developing countries, the appearance of most maturity indicators is delayed and that an adolescent takes longer to pass through the various stages of development, although the sequence of events is invariate. Conversely, urban children may experience maturation at equivalent or even earlier ages compared to their peers in developed countries (Cameron, 1993; Cameron et al., 1993).

The phenotypes of samples of adults in developing countries are characterized by mean heights that are usually below the 50th centile of NCHS references but are unlikely to be outside the range of heights to be found in developed countries. Weight and subcutaneous fat, on the other hand, tend to be relatively lower in males whilst females, particularly in Africa and South America, tend to have relatively greater weights and subcutaneous fat measurements.

Of importance to this discussion, however, is that, by five years of age, children in developing countries already demonstrate a lower growth status than their peers in the developed world (Martorell & Habicht, 1986). Some authors have suggested that this early loss of growth status in infancy permanently detracts from potential adult height. The evidence, however, is equivocal because of the difficulty of controlling for environmental factors between infancy and adulthood (Eveleth & Tanner, 1990). Certainly, improved nutrition and socioeconomic circumstances during childhood, in addition to the compensatory effect of adolescent growth, go some way towards diminishing such long-term effects (Cameron *et al.*, 1994). Ample evidence exists to record the fact that infants in developing countries begin to fall away from the NCHS references during the first two years of life (Cameron, 1991; Martorell & Habicht, 1986) but such is the stressful nature of the extrauterine environment that it is difficult or impossible to suggest that antenatal and birth factors *alone* account for this phenomenon. A poor growth status at five years of age, compared to the NCHS references, seems more likely to be the result of the postnatal environment, although there is no doubt that the antenatal and birth environments combine to affect the birthweight of the child. Even allowing for the homeorrhetic effect of postnatal catch-up growth (Prader, Tanner & von Harnack, 1963; Smith *et al.*, 1976), to bring the growth of a child to within its genetically predetermined growth canal (Waddington, 1957), there is little doubt that a child born of a low birthweight continues to manifest a growth disadvantage, compared to its normal birthweight peers, for a considerable period of time (Cruise, 1973). It is birthweight, therefore, that forms the initial indication as to whether there is likely to be a long-term consequence of an adverse antenatal environment.

Low birthweight

The cut-off point to delineate 'low birthweight' (LBW) has been internationally accepted as 2.5 kg, a weight that delineates 3–5% of the normal population in a developed country. Whilst this is a statistical cut-off point, children born below 2500 g are known to have a higher risk of morbidity and mortality during the first five years of life and particularly during the first year. Shapiro *et al.* (1980) found, for example, that LBW infants are almost 40 times more likely to die in the neonatal period than normal birthweight infants, and Starfield *et al.* (1982) noted that, whilst LBW babies represent only 6% of live births in the United States, they comprise 55% of infant deaths. Cruise (1973) described the growth of 202 low birthweight children in three gestational age categories; 28–32 weeks, 33–36 weeks, and 37–42 weeks. All three groups exhibited growth curves that were below those of children of normal gestational age and birthweight. The pre-term infants, however, exhibited catch-up growth surpassing the small-for-dates (SFD), or true LBW children, by one year. The maternal characteristics of this sample were that they were private patients and 'none suffered from economic deprivation'. In addition, the mothers had received antenatal medical care from obstetricians in private practice so it can be assumed that the antenatal factors that predispose a fetus to low birthweight were minimal.

Whilst a birthweight of 2.5 kg and less is internationally accepted as being 'low' and, in a developed country would prompt special monitoring and/or care in the perinatal period, data on the percentage of LBW children from a variety of developing countries in sub-Saharan Africa demonstrates that the combined mean birthweight is 3.02 kg and that 13.6% of children are LBW – three times greater than in the developed world (Table 5.1).

Results from birth-cohort studies in developing countries serve to illustrate the issues involved in investigating the relationship between growth and the antenatal and birth environments. The South African Birth To Ten study was initiated in response to the rapid growth of urban environments, and the subsequent realization that local knowledge of the health and growth of urban children was lacking (Yach *et al.*, 1991). The study *population* was composed of 5444 singleton births that took place between April 23rd and June 8th, 1990 to women who gave a permanent address within the metropolitan areas of Soweto and Johannesburg. From this population, 3770 cases (69.3%) were enrolled into the study during the first year. The majority of subjects were black (84.5%) or 'coloured' (7.3%) with fewer numbers of white (5.0%) or Indian (3.3%) cases. The external validity of the sample of black subjects was tested against registration records of all births occurring in the same time period and against births occurring in the two immediately preceding periods of seven weeks. The only significant difference was that enrolled mothers were significantly younger than unenrolled mothers (Chi^2 = 12.4; p < 0.01). Extensive information was collected on the mothers following enrolment at antenatal clinics (ANC) and delivery centres. Approximately 80% of pregnant women in Soweto attended ANCs by the thirtieth week of pregnancy. The questionnaire data related to socio-economic status, pregnancy history,

Table 5.1. Mean birth weights in Sub-Saharan Africa

	Birthweight (g)	%LBW
Southern Sahara		
Mali	3049	17
Chad	3114	11
Sudan	2966–3125	15
West Africa		
Senegal	3115	10
Guinea	2975	18
Ivory Coast	2950	14
Burkina Faso	2872	18
Nigeria	2880–3230	25
Cameroon	3119	13
Gabon	2979	16
Central Africa		
Congo	2907	12
Central African Rep.	2873	15
Zaire	3163	13
Uganda	3150	10
Rwanda	2890	17
Burundi	2730–3270	14
Angola	2876–3268	17
Zambia	2920–3180	14
East Africa		
Kenya	3143–3345	13
Tanzania	2906–3300	14
Southern Africa		
South Africa	3015–3300	12
Lesotho	3012	10
Madagascar	3145	10

housing conditions, and support during pregnancy. Following birth, the mothers and children were seen again at three months, and six months, and yearly thereafter. At each contact point, anthropometric and questionnaire data were collected. Failure to attend for assessment, or permanent attrition, naturally altered the sample size over the course of time. In addition, not all subjects completed all parts of the questionnaire on every occasion, thus samples for particular analyses tend to vary. Complete data on birthweight, gestational age, sex, and ethnicity were available on 2101 subjects who enrolled in the study either at the ANC or delivery data collection points.

Figure 5.2 illustrates the mean birthweights by population group and sex for those children of 'normal gestational age' (36–40 weeks). Mean birthweights are significantly different between sexes within population

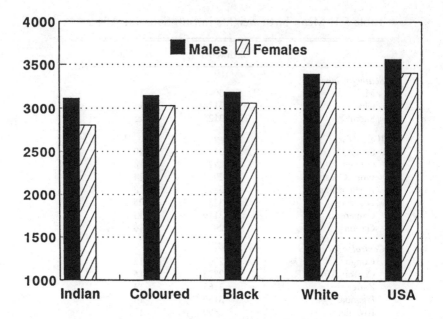

Figure 5.2. Mean birthweights of subjects of normal gestational age (36–40 weeks) in the Birth to Ten birth-cohort study in Soweto and Johannesburg, South Africa.

groups, but not between population groups. This lack of significance is most likely due to the small numbers of Indian and White subjects (15 and 16, respectively) compared to the black (1685) and 'coloured' groups (259). Ethnicity and birthweight are known to be significantly linked in the USA. Shiono *et al.* (1986) report that, compared to whites in the Northern California Kaiser-Permanente Birth Defects Study, relative mean differences among birthweights, adjusted for gestational age, for blacks was 183 g and for 'Asians' 167 g even after controlling for the joint effects of maternal smoking, alcohol use, sex of the child, parity, length of prenatal care, and maternal weight-for-height centile. Differences within the South African sample for blacks, 'coloureds' and Indians compared to whites were 243 g, 281 g, and 400 g, respectively. Such differences are usually explained by a differential access to antenatal health care, and there is no doubt that this contributes towards the lowered birthweights of these South African children. However, Shiono *et al.* (1986, p. 52) emphasize that their results of persistent differences in birthweight by ethnic group '... underscore the existing lack of understanding of the causes for (the) large ethnic differences...'. They question whether we are analysing appropriately the factors affecting the maternal environment such as socioeconomic status, physical and emotional stress, social support during pregnancy, and maternal nutrition. In the less subtle environment of South Africa, there is no doubt that ethnicity continues to play a major role. Perhaps surprisingly the percentages of LBW children among the South African black subjects were 7.6% compared to 12.5% amongst American blacks and were similar to the figure for northern Californian blacks of 7.7%.

The Birth To Ten study investigated the relationship between birthweight and 24 antenatal and delivery variables using one-way ANOVA on the three broad categories of maternal, social, and health variables. Only maternal age, parity, gravidity, and type of delivery, i.e. whether caesarian or normal, were significantly related to birthweight. The first three naturally group together because of the relationship of parity and gravidity to maternal age. Type of delivery becomes significant because of the preselection of greater birthweights amongst babies delivered by caesarian section. Cluster analysis to reduce the number of variables resulted in maternal age, maternal education, type of home and pelvic inflammatory disease as proxy measures of maternal, social and health factors, but controlling for maternal age resulted in a lack of significant effects for any of these variables on birthweight. Similarly, Shiono *et al.* (1986), using multiple linear regression, excluded the non-significant variables of gravidity, maternal age, education, religion, marital status, employment, previous miscarriage or stillbirth, anaemia, and hypertension. Statistically significant correlations with birthweight were found for pre-eclampsia, placental abruption, placenta previa, premature rupture of membranes, breech presentation, and type of delivery, but were excluded because they were *outcome* variables. In addition to ethnicity, significant effects on birthweight were found for smoking habit, sex for the child, parity, alcohol intake, maternal weight-for-height centile, and the timing of antenatal care.

Whilst Shiono *et al.* (1986) found no effect of maternal age, ample evidence exists to support the variables of maternal nutrition, age, and size as three of the most important antenatal factors affecting birthweight and postnatal growth. In addition, the macro-environmental characteristics of the mother, and particularly the altitude at which she lives, have an important bearing on birthweight and subsequent growth of children in the developing world.

Altitude

Huch (1986) estimates that between 20 and 30 million people live *permanently* above 3000 m and many of these are in the developing world. Bogin (1988) records that high altitude environments impose the stresses of hypoxia, high solar radiation, cold, low humidity, high winds and rough terrain with limitations of agricultural productivity. However, as with differences in mean birthweight by ethnic group, the relationship between altitude and mean birthweight is not a simple one.

Yip (1987) used the birth records from the United States natality files for all white infants born in the US from 1978 to 1981 to investigate the relationship between altitude and birthweight. 12.8 million individual records were used in the analysis allowing virtually ideal conditions for the creation of large random samples of singleton infants. These infants were compared across altitude gradations of 500 m except for the highest altitude range, which was grouped from 2500 to 3100 m. Sea level, or the reference sample, was the group born below 500 m. Two strategies were used for the comparison. In the first case, maternal factors of age, education and marital status were controlled within an analysis of covariance. In the second case, 'idealized' samples of infants from different altitudes were compared. These samples were selected to exclude premature infants and those from low socioeconomic status families. All infants were born to married women between 20 and 39 years of age; at least one parent had a college education and gestational age, calculated from last menstrual period, was equal to, or greater than, 37 weeks.

The birthweight reduction effect of altitude, using the data on the < 500 m group as the baseline, is statistically significant ($p < 0.001$). In addition, altitude was shown to be a strong predictor of low birthweight. Compared to infants born at sea level, infants born to women residing above 2500 m had a 10% reduction in mean birthweight and a *threefold* increase of low birthweight rate over the expected. This relative risk exceeds that of other known health and socioeconomic risk factors.

Whilst a variety of studies of samples living at altitude in South America have investigated the genetic and environmental correlates of low birthweight and the poor growth of high altitude children, it is not clear what physiological mechanism is responsible for birthweight reduction at altitude. Naturally, hypoxia is regarded as the most likely explanation. Moore *et al.* (1982*a,b*) provide evidence that low birthweight is associated with reduced maternal respiratory function and arterial oxygenation. It is not certain, however, whether *fetal* hypoxia is necessarily present *with* maternal hypoxia at high altitudes. Some human and animal studies (Cotter, Belechner & Prystowski, 1967; Sobrevilla *et al.*, 1967) have shown a similarity between fetal oxygenation levels at altitude *and* sea level. Yip (1987) believes that the most 'attractive' hypothesis is fetal physiological alteration related to maternal hypoxia.

Maternal nutrition

As recently as 1987, Prentice *et al.* (1987) were able to state that, with regard to the lower birthweights of children in the 'developing world' as compared to birthweights of children from industrialized nations, '... there is no consensus regarding the extent to which this deficit (in mean birthweights) is caused by acute maternal undernutrition...' (p. 912). Part of this confusion may result from the fact that the fetus is well protected from the effects of maternal malnutrition, and that this malnutrition must be present in the last few weeks of gestation for its effect to be observed clearly.

Smith's (1947) classic investigation of the relationship between maternal nutrition and birthweight of infants born in Holland towards the end of World War II, between 1944 and 1945, clearly demonstrated that the effect of maternal caloric intake on fetal weight was slight prior to the last trimester of pregnancy. The failure of the Allied airforces to capture bridgeheads across the great rivers at Nijmegen and Arnhem during the final phases of the war resulted in severe food shortages in Holland and led to famine during the 'hunger winter' of 1944–1945. The background was of general undernutrition rather than a shortage of selected foodstuffs. Birthweights were compared for infants born in Rotterdam during the winters of 1938–1939, 1943–1944, and 1944–1945. Using the former two groups as control groups, and with no significant change in maternal parity or age between the control periods and the hunger period, the net change in median birthweight was 240 g. At all centiles, the decline averaged 8% to 9% of the weight in normal times. Return to normal nutrition during the latter part of 1945 resulted in babies during that period who were an average of 200 g heavier than for both periods preceding the hunger winter. Beal (1971) who found no correlations between maternal intake of calories, carbohydrate, fat, protein and calcium and the length or weight of the infant at birth in a sample of white, middle-class American women, suggests that, when the prepregnancy nutrition of the mother has been good, few adverse effects are observed except for 'minor' decreases in birthweight. The longer and more severe the starvation, however, the more significant is the effect on both mother and infant.

Prentice *et al.* (1987) recorded the change in mean birthweights of the children of rural Gambian women over an eight-year period. The authors were interested in the effect of food supplementation during pregnancy on mothers who commonly have their first baby soon after menarche (mean age at first live birth = 18.4 yrs) and continue with regular reproductive cycles, of mean duration 29 months, until menopause. Even though birth spacing is extended by a prolonged period of lactational amenorrhea (mean = 13.5 months), sexual abstinence and contraception, pregnancy

Table 5.2.

	Sample size	Unadjusted mean (SEM)	Adjusted mean (SEM)
Wet season			
Presupplementation	48	2794 (57)	2842 (48)
Postsupplementation	106	3048 (38)	3031 (32)
Effect		+254 (68)***	+189 (62)*
Dry season			
Presupplementation	55	2891 (53)	2926 (45)
Postsupplementation	79	2950 (44)	2937 (37)
Effect		+59 (69)**	+11 (58)**
All year			
Presupplementation	103	2846 (39)	2882 (37)
Postsupplementation	185	3006 (29)	2961 (27)
Effect		+160 (49)**	+99 (44)**

***P<0.01 for pre-v postsupplementation.
*P<0.05 for pre-v postsupplementation.
**P<0.05 for supplementation in wet v dry season.
**P<0.05 for pre-v postsupplementation.
Adapted from a table in Prentice *et al.* (1987).
Mean = mean birthweight; SEM = sample error of the mean (g).

usually follows immediately after lactation, prior to nutritional recovery, and hence to nutritionally debilitated women.

Table 5.2 shows the analysis of birthweights of known gestational age both in their raw state and adjusted for sex, month of the year, parity and gestational age. The effect of supplementation was to increase mean birthweight by 254 (standard error of the mean (SEM)=68)g in the unadjusted data which was reduced to 189 (SEM=62)g after adjustment. Disaggregation of the data into wet and dry seasons demonstrated that the supplement was ineffective during the dry season, but highly effective during the wet season.

Research on the impact of maternal dietary supplementation on birthweight has found conflicting effects. These range from a positive effect through a small effect to no effect (Ashworth & Feachem, 1985; Allen, 1982; Adair & Pollitt, 1985; Thomson, 1983; Rush, 1982, 1983; Malhotra & Sawers, 1986). Studies in developing countries in South America (Columbia, Guatemala, Mexico, Chile) and Asia (Indonesia, Taiwan) cited by Prentice *et al.* (1987) all demonstrate positive effects of supplementation on birthweight. It is likely that maternal data must be disaggregated to reveal differences in terms of positive and negative energy balance, socioeconomic status and other confounding variables, before the benefits of supplementation become apparent. In addition, there may be a threshold of maternal energy intake below which food supplementation would be effective in improving birthweight (Prentice *et al.*, 1987; Susser, 1981, 1991). Such a threshold, about 1750 kcal/day according to Susser (1981), should not be interpreted as critical but rather as an indication that birthweight is reduced when women are unable to maintain a positive energy balance. Obviously, the latter itself is dependent on prenatal energy reserves, body size and the level of physical activity. Among the majority of women in developing countries, small body size due to chronic malnutrition during childhood, low energy reserves due to close birth spacing, and relatively high levels of physical activity throughout pregnancy, suggest that food supplementation would be effective in increasing birthweight and decreasing the percentage of LBW children. In addition, Prentice *et al.* (1987) calculated the possible effect of food supplementation on the infant mortality rate which had already been reduced to 45/1000 live births by the presence of their clinic in Keneba. They predicted, on the basis of increased birthweight, that a 37% reduction in neonatal mortality would result from dietary supplementation.

Maternal age

There is little doubt that teenage pregnancy is associated with an increased incidence of low birthweight children particularly in developing countries. In the Nairobi birth survey (NBS) between June and August 1981, for example, Bwibo (1985) recorded a mean birthweight of 2920 (SD = 553) g for the singleton offspring of 944 teenage mothers, the youngest of whom was 10 yrs 2 months. In contrast, the mean birthweight of 3133 g (SD = 533) was recorded for 5293 singleton offspring in the general population. The percentage of LBW was 14% and 7%, respectively. From a more extensive study in Nigeria, Bwibo (1985) quotes LBW figures of 27% for mothers less than 15 years, 26% in mothers 15 to 19 years, 20% in mothers 20 to 24 years and 18% in mothers 25 to 29 years.

Both maternal nutrition and maternal size are related intimately in the adolescent mother who, in addition to maintaining the growth of the fetus, must also maintain her own growth. Studies to investigate the cost of pregnancy to the adolescent mother's physical growth are few (Garn *et al.*, 1984; Sukanich, Rogers & McDonald, 1986) because of the lack of good instrumentation to measure short-term growth, and the fact that statural changes during pregnancy may be affected by postural changes, weight gain and compression of the intervertebral discs (Scholl *et al.*, 1988). Both Garn *et al.* (1984) and Sukanich *et al.* (1986) report increments in height and weight between successive pregnancies in well-nourished American teenage girls. However, Garn *et al.* (1984) emphasize that the weight gains they measured were generally less than would be expected for the maturational level of their subjects. Scholl *et al.* (1988), using a knee-height measuring device, purport to be the first to demonstrate that pregnant adolescents, enrolled in an adolescent pregnancy study in Camden, New Jersey, continue to grow *during* pregnancy. Twenty-one primagravidae, aged 12 to 15 years, gained 1.68 mm as compared to 24 multigravidae, aged 15 to 18 years (with a first pregnancy when aged 12 to 15 years), who grew 1.00 mm over a 14-week period during the second and third trimester. Nine mature controls aged 18 to 29 years experienced an average of 0.10 mm increase in the same period. The assumption is made that these American adolescents received adequate nutrition. That assumption cannot be made within the scenario of a developing country. It is more likely that pregnancy combines with chronically low nutritional intake to compromise the growth during pregnancy of the adolescent gravida on such countries.

Relationships between birthweight and older mothers would appear to be more complex. This may be due to the fact that the child born to an older mother is likely to have greater parity and thus be subjected to different influences from the offspring of the adolescent mother. Selvin & Janerich's (1971) study demonstrates this added complexity. The first child born to a mother older than 24 years was consistently lighter than that born to mothers aged 20 to 24 years. However, children of higher birth order were consistently heavier than 'only children' even though they continued to demonstrate a relationship between maternal age and birthweight. It is not until infants with a parity of 6 or greater are born that birthweight consistently increases with increasing maternal age. No data were supplied for the youngest maternal age group because only four mothers had produced six or more children by the age of 20 years!

Parity therefore combines with maternal age to produce a complex picture. Thomson, Billiwicz & Hytten (1968) demonstrated, from the British Perinatal Mortality Survey, that second and subsequent babies grow faster than first babies and Tanner & Thomson (1970) developed birthweight charts to relate parity with the normality of birthweight. Firstborn children are naturally constrained in the growth during the final few weeks of gestation because of the mechanical constraints of the uterus. Once the uterus has been effectively stretched by the first child, subsequent children have a larger space within which to grow.

Maternal size

Maternal size is well recognized as being associated with birthweight and is recognized as a significant factor in secular trend studies (Alberman *et al.*,

1991). Various studies have shown regression coefficients of 12.1 to 22.1 g/cm when height alone is considered as an independent regression variable against birthweight, and 12.5 to 20.0 g/kg when weight is entered independently into such a regression (Carr-Hill & Pritchard, 1985; Thomson, 1983; Love & Kinch, 1965; McKeown & Record, 1954; Briend, 1985; O'Sullivan *et al.*, 1965). Such effects are to be observed most easily in the absence of confounding factors such as maternal nutrition. When maternal weight gain is taken into account, a more complex picture emerges both in terms of absolute nutritional intake, as discussed earlier, and in terms of the timing of the poor intake.

Maternal weight gain is important in terms of both birthweight and length of gestation, but the pattern of weight gain also plays a role. Early and late weight gains during pregnancy have independent effects on pregnancy outcome (Hediger *et al.*, 1989). During adolescent pregnancy, inadequate weight gains are associated with increased risk of small for gestational age babies. Importantly, this risk is not diminished by later adequate nutrition or compensatory gains to levels recommended for adult gravida. Such inadequate nutrition during early pregnancy assumes significance in developed countries where the adolescent gravida may respond to body image concerns and in developing countries where chronic undernutrition, particularly with regard to energy intake, is endemic. The mechanism for reducing fetal growth is likely to be through interference with early placental development, limiting transfer of nutrients to the fetus which restricts fetal growth, even though later nutrition is adequate.

Walker & Stein (1985) have suggested that LBW may be associated with the smaller size of the mother in developing countries. Prentice *et al.* (1987, p. 915) performed a multiple regression analysis of birthweight against the maternal variables of height, weight and weight gain during pregnancy. Variables were entered independently and in combination (height + weight; height + weight gain; weight + weight gain, height + weight + weight gain). Prentice *et al.* (1987) found that birthweight increased by 17.6 g/cm when maternal height alone was considered as an independent regression variable. When maternal weight was entered independently into the regression, it had a coefficient of 14.0 g/kg. Height, weight and maternal weight gain, when entered simultaneously, reduced these coefficients to 11.6 g/cm and 6.2 g/kg, respectively. Adjustments to mean birthweights are, therefore, possible to partition out the effects of small maternal size and gain a clearer impression of the effect of other factors. When Prentice *et al.* (1987) accomplished this, they statistically accounted for an increase in mean birthweight by 84 g (20 g from the smaller heights of the mothers and 64 g from the smaller weights), and were left with an unexplained difference between Gambian birthweights and British birthweights of 416 g. Much of this, they suggested, was caused by acute maternal malnutrition during pregnancy.

There is little doubt, therefore, that some of the reduction in birthweight and the subsequent increase in the percentage of LBW is due to the absolutely smaller size of women in developing countries. That smaller size is, in part, due to the early nutritional deprivation experienced by the vast majority of such women. As an example of this effect on birthweight, if one takes the range of mean heights of adult African women illustrated by Eveleth & Tanner (1976), the majority cluster between 153 and 160 cm with an average around 156 cm. If one considers that the average height of British women was 162 cm, height alone, at an average coefficient of 15 g/cm, would reduce birthweight by about 90 g compared to the birthweights of British children. A similar calculation using maternal weight would reduce birthweight by 75 g. It could be suggested therefore, that African birthweights are low by 165 g simply because of smaller maternal size and the residual is due mainly to maternal nutrition.

Conclusions

This review of the factors affecting the antenatal and birth environments, and their subsequent effect on the growth and phenotype of the offspring, has used birthweight as the major outcome variable and concentrated on maternal factors that influence this outcome. The social milieu within which the child is conceived, born, and raised, has not been reviewed because the causative association between the complex array of variables that combine to create a social class effect and their actual dynamic with human growth has still not been clarified. In part, this failure is due to current inability to measure the actual variables that interact to produce the pattern of growth, both prenatally and postnatally, that is associated with differences in social class. Instead, proxy measures of parental occupation, income, and educational achievement are used without really understanding how such variables actually affect the child.

It is obvious that the extrauterine environment immediately begins to affect the child following birth and that, to a large extent, the child born of low, but not *clinically* low, birthweight can recover from many of the adverse factors affecting the mother if he/she is subjected to an improved environment. It is generally accepted that secular trends for increasing birthweights, better growth status, earlier maturation, and greater adult statures are indicators of improved environmental conditions affecting the mother during pregnancy and her children during growth. It may be, however, that the long-term effects of adverse conditions *in utero* persist to affect not only the adult phenotype and disease profile but also the phenotypic variance of the next generation.

References

Adair, L.S. & Pollitt, E. (1985). Outcome of maternal nutritional supplementation: a comprehensive review of the Bacon Chow Study. *American Journal of Clinical Nutrition*, **41**, 948–78.

Alberman, E., Filakti, H., Williams, S. & Evans, S.J.W. (1991). Early influences on the secular change in adult height between the parents and children of the 1958 birth cohort. *Annals of Human Biology*, **18**, 127–36.

Alberman, E., Emanuel, I., Filakti, H. & Evans, S.J. (1992). The contrasting effects of parental birthweight and gestational age on the birthweight of offspring. *Paediatric and Perinatal Epidemiology*, **6**, 134–44.

Allen, L.H. (1982). Provision of nutrient supplements to pregnant women. *Nutrition Research*, **2**, 115–16.

Ashworth, A. & Feachem, R.G. (1985). Interventions for the control of diarrhoeal diseases among young children: prevention of low birthweight. *Bulletin of the World Health Organisation*, **63**, 165–84.

Barker, D.J.P., Osmond, C., Golding, J., Kuh, D. & Wadsworth, M.E.J. (1989). Growth *in utero*, blood pressure in childhood and adult life, and mortality from cardiovascular disease. *British Medical Journal*, **298**, 564–7.

Beal, V.A. (1971). Nutritional studies during pregnancy. II. Dietary intake, maternal weight gain and size of the infant. *Journal of the American Dietetic Association*, **58**, 321–33.

Bogin, B. (1988). *Patterns of Human Growth*. Cambridge: Cambridge University Press.

Briend, A. (1985). Do maternal energy reserves limit fetal growth? *Lancet*, **i**, 38–40.

Bwibo, N.O. (1985). Birthweights of infants of teenage mothers in Nairobi. *Acta Paediatrica Scandinavica* (Suppl.), **319**, 89–94.

Cameron, N. (1986). Standards for human growth: their construction and use. *South African Medical Journal*, **70**, 422–5.

Cameron, N. (1991). Human growth, nutrition, and health status in sub-Saharan Africa. *Yearbook of Physical Anthropology*, **34**, 211–50.

Cameron, N. (1993). Assessment of growth and maturation during adolescence. *Hormone Research*, **39** (Suppl. 3), 9–17.

Cameron, N., Gordon-Larsen, P. & Wrchota, E.M. (1994). Longitudinal analysis of adolescent growth in height, fatness, and fat patterning in rural South African black children. *American Journal of Physical Anthropology*, **93**, 307–21.

Cameron, N., Grieve, C.A., Kruger, A. & Leschner, K.F. (1993). Secondary sexual development in rural and urban South African black children. *Annals of Human Biology*, **20**, 583–93.

Carr-Hill, R. & Pritchard, C. (1985). *The Development and Exploitation of Empirical Birthweight Standards*. New York: Stockton Press.

Cole, T. (1993). The use and construction of anthropometric growth reference standards. *Nutrition Research Reviews*, **6**, 19–50.

Cotter, J.R., Belechner, J.N. & Prystowski, H. (1967). Observations on pregnancy at altitude. I. The respiratory gases in maternal arterial and uterine venous blood. *American Journal of Obstetrics and Gynecology*, **99**, 1–8.

Cruise, M.O. (1973). A longitudinal study of the growth of low birth weight infants. 1. Velocity and distance growth, birth to 3 years. *Pediatrics*, **51**, 620–8.

Dettwyler, K.A. & Fishman, C. (1992). Infant feeding practices and growth. *Annual Review of Anthropology*, **21**, 171–204.

Dubos, R., Savage, D. & Schaedler, R. (1966). Biological Freudianism: Lasting effects of early environmental influences. *Pediatrics*, **38**, 789–800.

Eisner, V., Brazie, J.V., Pratt, M.W. & Hexter, A.C. (1979). The risk of low birthweight. *American Journal of Public Health*, **69**, 887–93.

Eveleth, P.B. & Tanner, J.M. (1976). *Worldwide Variation in Human Growth*, 1st edn. Cambridge: Cambridge University Press.

Eveleth, P.B. & Tanner, J.M. (1990). *Worldwide Variation in Human Growth*, 2nd edn. Cambridge: Cambridge University Press.

Garn, S.M., La Velle, M., Pesick, S.D. & Ridella, S.A. (1984). Are pregnant teenagers still in rapid growth? *American Journal of Diseases in Children*, **138**, 32–4.

Hamill, P.V.V., Drizd, T.A., Johnson, C.L., Reed, R.B., Roche, A.F. & Moore, W.M. (1979). Physical growth: National Center for Health Statistics percentiles. *American Journal of Clinical Nutrition*, **32**, 607–29.

Hediger, M., Scholl, T.O., Belsky, H., Ances, I.G. & Salmon, R.W. (1989). Patterns of weight gain in adolescent pregnancy: effects on birth weight and preterm delivery. *Obstetrics and Gynecology*, **4**, 6–11.

Heikkinen, T., Alvesalo, L., Osborne, R.H. & Pirttiniemi, P. (1992). Maternal smoking and tooth formation in the foetus. *Early Human Development*, **30**, 49–59.

Huch, R. (1986). Maternal hyperventilation and the fetus. *Journal of Perinatal Medicine*, **14**, 3–17.

Kleinman, J.C. & Kessel, S.S. (1987). Racial differences in low birth weight. *The New England Journal of Medicine*, **317**, 749–53.

Law, C.M., Barker, D.J., Osmond, C., Fall, C.H. & Simmons, S.J. (1992). Early growth and abdominal fatness in adult life. *Journal of Epidemiology and Community Health*, **46**, 184–6.

Leff, M., Orleans, M., Haverkamp, A.D., Bar-on, A.E., Alderman, B.W. & Freedman, W.L. (1992). The association of maternal low birthweight and infant low birthweight in a racially mixed population. *Peadiatric and Perinatal Epidemiology*, **6**, 51–61.

Love, E.J. & Kinch, R.A.H. (1965). Factors influencing birthweight in normal pregnancy. *American Journal of Obstetrics and Gynecology*, **91**, 342–9.

Lumey, L.H. (1992). Decreased birthweights in infants after maternal *in utero* exposure to the Dutch famine of 1944–1945. *Paediatric and Perinatal Epidemiology*, **6**, 240–53.

McKeown, T. & Record, R.G. (1954). Influence of prenatal environment on correlation between birthweight and parental height. *American Journal of Human Genetics*, **6**, 457–63.

Malhotra, A. & Sawers, R.S. (1986). Dietary supplementation in pregnancy. *British Medical Journal*, **293**, 465–6.

Martorell, R. & Habicht, J-P. (1986). Growth in childhood in developing countries. In *Human Growth: A Comprehensive Treatise*, ed. F. Falkner & J.M. Tanner

2nd edn., pp. 241–62. New York: Plenum.

Moore, L.G., Jahnigen, D., Rounds, S.S., Reeves, J.T. & Grover, R.F. (1982*a*). Maternal hyperventilation helps preserve arterial oxygenation during high altitude pregnancy. *Journal of Applied Physiology*, **52**, 690–4.

Moore, L.G., Rounds, S.S., Jahnigen, D., Grover, R.F. & Reeves, J.T. (1982*b*). Infant birth weight is related to maternal arterial oxygenation at high altitude. *Journal of Applied Physiology*, **52**, 695–9.

Mueller, W.H. (1986). The genetics of size and shape in children and adults. In *Human Growth: A Comprehensive Treatise*, ed. F. Falkner and J.M. Tanner, 2nd edn, vol. 3, pp. 145–68. New York: Plenum.

O'Sullivan, J.B., Gellis, S.S., Tenny, B.O. & Mahan, C.M. (1965). Aspects of birthweight and its influencing variables. *American Journal of Obstetrics and Gynecology*, **92**, 1023–9.

Prader, A., Tanner, J.M. & von Harnack, G.A. (1963). Catch-up growth following illness or starvation. *Journal of Pediatrics*, **62**, 646–59.

Prentice, A.M., Cole, T.J., Foord, F.A., Lamb, W.H. & Whitehead, R.G. (1987). Increased birthweight after prenatal dietary supplementation of rural African women. *American Journal of Clinical Nutrition*, **46**, 912–25.

Roche, A.F. (1992). *Growth Maturation and Body Composition: The Fels Longitudinal Study 1929–1991*. Cambridge: Cambridge University Press.

Rush, D. (1982). Effects of changes in protein and calorie intake during pregnancy on the growth of the human fetus. In *Effectiveness and Satisfaction in Antenatal Care*, ed. M. Enkin and I. Chalmers, pp. 92–113. Philadelphia: JB Lippincott.

Rush, D. (1983). Effects of protein and calorie supplementation during pregnancy on the fetus and developing child. In *Nutrition in Pregnancy*, ed. D.M. Campbell and M.D.G. Gillmer, pp. 65–81. London: Royal College Obstetricians and Gynaecologists.

Scholl, T.O., Hediger, M.L., Ances, I.G. & Cronk, C.E. (1988). Growth during early teenage pregnancy. *Lancet*, **i**, 701–2.

Schreuder, A.M., Veen, S., Ens-Dokkum, M.H., Verloove-Vanhorick, S.R., Brand, R. & Ruys, J.H. (1992). Standardised method of follow-up assessment of preterm infants at the age of 5 years: use of WHO classification of impairments, disabilities and handicaps. Report from the Collaborative Project on Preterm and Small for Gestational Age Infants (POPS) in the Netherlands, 1983. *Paediatric and Perinatal Epidemiology*, **6**, 363–80.

Selvin, S. & Janerich, D.T. (1971). Four factors influencing birthweight. *British Journal of Preventive and Social Medicine*, **25**, 12–20.

Shapiro, S., McCormick, M., Starfield, B., Krischer, J. & Bross, D. (1980). Relevance of correlates of infant deaths for significant morbidity at 1 year of age. *American Journal of Obstetrics and Gynecology*, **136**, 363–73.

Shiono, P.H., Klebanoff, M.A., Graubard, B.I., Berendes H.W. & Rhoads, G.G. (1986). Birth weight among women of different ethnic groups. *Journal of the American Medical Association*, **255**, 48–52.

Silva, P.A. (1990). The Dunedin Multidisciplinary Health and Development Study: a 15 year longitudinal study. *Paediatric and Perinatal Epidemiology*, **4**, 76–107.

Smith, C.A. (1947). Effects of maternal undernutrition upon newborn infants in Holland (1944–1945). *Journal of Pediatrics*, **30**, 229–41.

Smith, P.W., Truog, W., Gogers, J.E., Greitzer, L.J., Skinner, A.L., McCann, J.J. & Harvey, M.A.S. (1976). Shifting linear growth during infancy: illustration of genetic factors in growth from foetal life through infancy. *Journal of Pediatrics*, **89**, 225–30.

Sobrevilla, L.A., Romero, Q.F.I., Kruger, F. & Whittembury, J. (1967). Low estrogen excretion during pregnancy at high altitude. *American Journal of Obstetrics and Gynecology*, **102**, 828–33.

Starfield, B., Shapiro, S., McCormick, M. & Ross, D. (1982). Mortality and morbidity in infants with intrauterine growth retardation. *Journal of Pediatrics*, **101**, 978–83.

Sukanich, A.C., Rogers, K.D. & McDonald, H.M. (1986). Physical maturity and outcome of pregnancy in primaparas younger than 16 years of age. *Pediatrics*, **78**, 31–6.

Susser, M. (1981). Prenatal nutrition, birthweight, and psychological development: an overview of experiments, quasi-experiments and natural experiments in the last decade. *American Journal of Clinical Nutrition*, **34**, 784–803.

Susser, M. (1991). Maternal weight gain, infant birth weight, and diet: causal sequences. *American Journal of Clinical Nutrition*, **53**, 1384–96.

Suzanne, C. (1980). Developmental genetics of man. In *Human Physical Growth and Maturation*, ed. F.E. Johnston, A.F. Roche and C. Suzanne, pp. 221–42. New York: Plenum.

Tanner, J.M. & Thomson, A.M. (1970). Standards for birthweight at gestation periods from 32 to 42 weeks, allowing for maternal height and weight. *Archives of Disease in Childhood*, **45**, 566–77.

Thomson, A.M. (1983). Fetal growth and size at birth. In *Obstetrical Epidemiology*, ed. S.L. Barron and A.M. Thomson, pp. 89–142. London: Academic Press.

Thomson, A.M., Billiwicz, W.Z. & Hytten, F.E. (1968). The assessment of fetal growth. *Journal of Obstetrics and Gynecology*, **75**, 903–9.

Waddington, C.H. (1957). *The Strategy of Genes*. London: Allen & Unwin.

Walker, A.R.P. & Stein, H. (1985). Growth of third world children. In *Dietary Fibre, Fibre Depleted Foods and Disease*, ed. D.P. Birkett, H.C. Trowell and K.J. Heaton, pp. 331–44. London: Academic Press.

Werner, E.E. & Smith, R.S. (1982). *Vulnerable but Invincible*. New York: McGraw-Hill.

Wright, C.M., Waterson, A. & Aynsley-Green, A. (1993). Comparison of the use of Tanner and Whitehouse, NCHS, and Cambridge standards in infancy. *Archives of Diseases in Childhood*, **69**, 420–2.

Yach, D., Cameron, N., Padayachee, N., Wagstaff, L., Richter, L. & Fonn, S. (1991). Birth To Ten: Child health in South Africa in the 1990s. Rationale and methods of a birth cohort study. *Paediatric and Perinatal Epidemiology*, **5**, 1–23.

Yip, R. (1987). Altitude and birth weight. *Journal of Pediatrics*, **X**, 869–76.

6 *The effect of early nutrition on later growth*

MICHAEL H.N. GOLDEN

Introduction

Underprivileged children are, on average, shorter than those nurtured in affluent surroundings. About one-third of the world's children are less than two standard deviations below North American standards; this stunting, which starts in infancy or earlier, usually persists to give rise to a small adult. The first question to be addressed is whether an adverse early nutritional environment permanently affects the individual to give a 'growth scar' that persists to adult life or whether catch-up can occur if conducive conditions are provided.

Martorell *et al.* (1979, 1990) examined the increment in height of stunted and non-stunted Guatemalan children between 5 and 18 years of age. The height gain during this period was greater than that of US citizens of Mexican ancestry and only 3 centimetres less than NCHS standards; in other words, the absolute gain in height was 'normal'. As the increment was totally independent of the subjects' heights at age 5 years, the initially stunted children remaining short throughout their subsequent growth, Martorell *et al.* concluded that stunting is, 'a condition resulting from events in early childhood and which, once present, remains for life.... There is no catch-up growth in later childhood and adolescence as some might have expected'. This conclusion was reinforced by the very high correlation they found between the height of children at 3 and their final adult height and performance (Martorell *et al.*, 1992). Such a high correlation between height in childhood and final adult height is well known in other populations from both the developed and developing world (Mills *et al.*, 1986; Satyanarayana, Prasanna-Krishna & Narasinga-Rao, 1986; Binkin *et al.*, 1988). These data seem to show that a period of malnutrition in the first years of childhood irrevocably changes the child so that they are 'locked into' a growth trajectory with a lower potential for final adult height. The growth studies of Karlberg (Karlberg, 1989) in

91

which growth is divided, largely by statistical deconvolution, into an infantile, a childhood and a pubertal component provide a potential mechanism for these observations; if the infantile stage is abnormal or the onset of the childhood component is delayed and shortened, the stunting may be irreversible because these stages cannot be re-entered; this seems to be supported by studies of growth of children in Pakistan (Karlberg, Jalil & Lindblad, 1988).

However, if there are populations of stunted children that do catch up substantially in height, to reach normal adult values, the hypothesis of Martorell (Martorell *et al.*, 1990) must be amended or rejected. The alternative hypotheses are (i) that full catch-up growth is indeed possible but does not often occur; this could be because the correct conditions for catch-up are not satisfied owing to an unchanging environment and diet that initially led to poor growth performance; (ii) that the insult suffered by the children studied by Martorell is qualitatively different in type, severity or timing from children that do catch up in height; in this case we need to understand why some children or populations are affected permanently whilst others are not; or (iii) that true 'catch-up' does not occur when one of the phases of growth have been compromised, but that the deficit can be compensated by additional growth in one of the other phases.

Can children that are stunted in early childhood catch up in height?

Evidence that stunted children can indeed catch up in height comes from several sources: cross-sectional studies of stature at different ages in underprivileged communities, and longitudinal observations of children that are stunted in early childhood who either catch up spontaneously, change their environment, or have chronic disease treated.

Compelling data to show that remarkable catch-up is possible even under conditions of extreme privation is presented by Steckel (1987). Following the abolition of the American slave trade from Africa in 1807, all slaves transported by sea from one US port to another had to have their names, ages, sex and heights recorded on the ship's manifest. Steckel analysed the heights of 50 606 such slaves. The recorded heights presented as centiles of present-day (NCHS) standards are shown in Figure 6.1. The children are tiny by modern standards, as small as the Bundi in Papua New Guinea studied by Malcolm; they remain small throughout childhood and as juveniles with very little change in their centiles; however, between 15 and 17 years of age they quite suddenly undergo a spectacular catch-up in height to above the modern 25th centile for males and the 35th centile for females. At the time of catch-up they receive extra food, although they start exhausting work in the fields. These data clearly show that adolescents can have an almost complete catch-up in height without any change in hygiene or exposure to disease.

In Chile (Alvear *et al.*, 1986) follow-up, for about 6 years, of malnourished children that had returned to a very poor environment showed clear evidence of spontaneous catch-up (87 to 92% of standard); these children not only had postnatal malnutrition but many also had a low birthweight and their mothers were much smaller than local controls, they were the most underprivileged of a poor society.

Graham's group (Graham & Adrianzen, 1971) studied very stunted malnourished children in Peru. Their height–age was only 45% of chronological age at 1 year of age; by 3 years they had caught up to 61% and by 7 years were 69%. Later studies showed continued catch-up (Graham *et al.*, 1982); however, the girls did decidedly better than the boys (girls were 2 cm shorter than boys at 1 year and 7 cm taller by 13 years). Remarkably, the girls were taller than their own mothers from age 14 years onwards: the boys lagged behind their fathers at age 16 (150 v 158 cm). It is clear that the girls had a very marked catch-up, probably to their height potential. The same gender difference is seen in Steckel's data from the slaves where the girls do decidedly better than the boys. One plausible reason for the difference in the girls and boys is that the requirement of zinc for growth is very much higher in males than females, at all ages from *in utero* to adulthood, and in all mammalian species so far investigated. The major feature of zinc deficiency is retarded growth (Golden, 1988); in most studies of zinc supplementation only the males respond (Walravens, Hambidge & Koepfer, 1989) zinc limitation could easily account for the failure of the males to catch up in Peru; it is clear that zinc, and the other 'growth' or type II nutrients (Golden, 1991), have been ignored and usually remain deficient during the later growth of poor children in the USA as well as in the developing world (Hambidge, 1986; Walravens *et al.*, 1983, 1989; Walravens & Hambidge, 1976). One wonders what would have been the final height of the slaves in nineteenth century America, Peru, Chile or even the Guatemalans studied by Martorell *et al.*, if type II nutrient rich food had been regularly taken.

In Cape Town (Keet *et al.*, 1971) only 8% of malnourished children were above the third centile of the Boston growth standard on admission, at 5- and 10-years follow-up 22 and 42%, respectively, were above this cut-off point and virtually all individuals were in a higher centile group at the 10-year than at 5-year follow up. This catch-up by older children is notable because there was no intervention and the children had a high incidence of hypoalbuminaemia at follow-up, so that they caught up despite continuing malnutrition. At the 15-year follow-up the catch-up had continued with

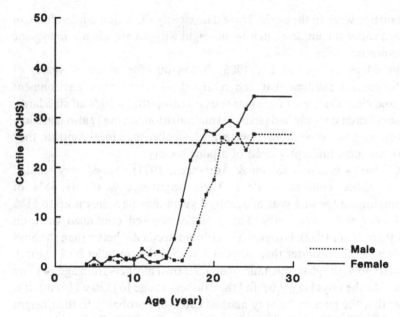

Figure 6.1. Heights of slaves transported by sea from one part of the USA to another between 1820 and 1860, by age. The data are plotted as centiles of the NCHS standard, the horizontal line is at the 25th centile. Data from Steckel (1987).

annual increments in height above those expected (Bowie *et al.*, 1980). These findings are in marked contrast to those reported from Guatemala. Interestingly, both the boys and the girls were continuing to gain substantial amounts of height after 17 years when the standards approach an asymptote. The final adult height of these subjects has not been reported.

In Kenya, Kulin *et al.* (1982) conducted a cross-sectional survey comparing girls and boys from three privileged schools in Nairobi with an impoverished rural district with a very prevalence of malnutrition. Table 6.1 gives a summary of their height data. Before puberty, the rural children were 1.7 to 2.0 standard deviations behind the well-nourished children. By the age of 18 years, the previously malnourished rural children had completely caught up with the affluent children, although puberty was delayed by 2 to 3 years in the malnourished group. This is not a unique finding; malnourished and well-nourished children who reach approximately the same final adult height were reported almost 30 years ago (Dreizen, Spirakis & Stone, 1967).

Cross-sectional measurements of the Turkana pastoralists of Northern Kenya show that by the age of 13 they are almost 15 cm behind the NCHS

Table 6.1. *Height (cm ± SD) of school children from rich (urban) and poor (rural) schools in Kenya. Data of Kulin et al. (Kulin et al., 1982)*

Age	Male				Female			
	Age 10		Age 18		Age 10		Age 18	
Well nourished	142.9	±7.4	164.4	±6.0	144.3	±5.9	156.9	±7.3
Malnourished	130.4	±5.8	164.8	±5.0	123.7	±8.0	157.3	±5.9
Difference (cm)	−12.5		+0.4		−20.6		+0.4	

standards (Little, Galvin & Mugambi, 1983), but by the time they are adults they have completely caught up.

The close correlation between childhood height and adult height implies that there is a deficit that remains constant throughout growth. The children from Guatemala seem to demonstrate this phenomenon (Martorell *et al.*, 1990, 1992). However, examination of the growth of the Bundi children from Papua New Guinea shows that they are about 15 cm shorter than the standards at age 2, and 20 cm shorter at adulthood (Malcolm, 1971). However, they become steadily more stunted throughout childhood so that they are almost 40 cm behind the standards at 12–14 years of age; they then catch up by about 20 cm after puberty.

The growth trajectories from these studies are not compatible with the simplistic view that there is a lost period of growth which cannot be made good after early childhood. Although the Turkana and Bundi children could be interpreted as demonstrating compensatory growth during the pubertal phase it should be remembered that the genetically determined growth pattern of these particular groups might be quite different from the population (USA) giving rise to the standards.

Although it seems that the data from Africa are qualitatively different from those derived from Latin America, there are studies from Africa which show a total failure to spontaneously catch up, for example, Pelletier, Low & Msukwa (1991).

Studies from India show a pattern similar to that seen in Peru. There is a great deal of stunting in early childhood, the girls catch up somewhat (Satyanarayana *et al.*, 1981), whereas the boys maintain their deficit (Satyanarayana, Nadamuni & Narasinga Rao, 1980), a pattern typical of chronic zinc deficiency. Nevertheless, the most stunted boys do catch up relative to the less stunted boys in the same cohort (Satyanarayana *et al.*, 1989). Again, it must be stressed that in none of the studies was there any intervention and the children usually remained grossly deprived in many ways (Satyanarayana *et al.*, 1986).

These studies of malnourished children in poor environments show that spontaneous catch-up is possible without an environmental change; what happens when there is a change in circumstances?

Deprivation also occurs in affluent societies. King & Taitz (1985) studied 96 abused children in England, 64 of whom were discharged home and 31 received either foster or mixed care. There was a considerable catch-up in the children's height which was more marked in the children that were fostered.

Turkish children born in Turkey are shorter than those born in Sweden when they arrive in the country but they quickly catch up (Mjones, 1987) in their new environment; similarly, follow-up of immigrants to a poor area of Glasgow from Africa, Asia and China showed the immigrants all more advanced in height and bone age than their Scottish counterparts (Goel *et al.*, 1981).

A longitudinal study of 835 refugee children entering the USA was reported by Schumacher, Paulson and Kretchmer (1987); they were analysed in four racial groups and seven age cohorts, by sex. On arrival, nearly all groups were below the 25th NCHS centile; they gained height at an accelerated rate, frequently achieving mean growth rates near the 90th centile for US children. There was no indication that the younger children cohorts had relatively greater rates of gain than the older children.

Graham's group (Graham & Adrianzen, 1971) conducted a unique experiment. They kept children from very poor families in a convalescent hospital for 18 months and compared their growth with that of siblings at home. At discharge, the hospital children had a height of -1.1 Z; the children at home were -2.3 Z. Five of the children remain in the hospital and 13 were discharged; nine months later the hospital group were still -1 Z whereas those returned home were now -2.2 Z. However, in the long term, giving these infants a 'good start' did not help for, after discharge, their growth curve from 2 to 8 years of age was almost identical to the children that had not been kept in hospital (Baertl, Adrianzen & Graham, 1976).

Graham also examined the opposite effect of going from a poor to a good environment (Graham & Adrianzen, 1972). There was a growth retardation of about -3.5 Z in two groups of eight children, aged 8 and 9 months on admission to hospital. At 15 months the control group were discharged to their parents; they had caught up in hospital to -2.9 Z; at about 9 years of age they were still grossly stunted (-2.5 Z). The second group of children were kept in hospital longer, to age 26 months, and caught up more to -2.2 Z. They were discharged home, however, at 4 years of age, when still severely stunted (-2.1 Z), they were adopted. In contrast to the children that returned home, these older adoptees almost caught up to modern American standards ($-0.5\,Z$) and far exceeded the local standards. This study shows how children malnourished during the first few years of life can make a complete recovery before puberty by changing their home environment within the Third World. Unfortunately, these children have not yet been followed to adulthood (Graham, G.G., personal communication).

Winick, Meyer and Harris (1975) and Lien, Meyer and Winick (1977) studied female Korean children that had been adopted by families in the USA when aged 2 to 5 years. They were re-measured 4 to 15 years later by which time they had all caught up to Korean standards, the taller adoptees reaching well above the Korean median.

A similar study has been conducted by Proos, Hofrander and Tuvemo (1991a). They studied Indian children adopted by Swedish parents. Half the children were between the ages of 3 and 6 years. The children had a very pronounced catch-up within 2 years of arrival in Sweden. Those with the largest deficits caught up the most ($r = 0.53$). By the age of menarche, they had caught up further to now be only $-0.3\,Z$ below the NCHS standard. Thus, by the time of puberty, there had indeed been an almost complete catch-up to a level well above local Indian standards. However, they had an early puberty (Proos, Hofrander & Tuvemo, 1991b), and the increment in growth over puberty was the same as that of Swedish girls so that their final adult stature was $-1.4\,Z$ with 31% of the children being more than $2\,Z$ below the NCHS standard. There was a close relationship between the height when adopted and their final adult height ($r = 0.78$). The absolute increment from adoption to adulthood is not reported, but it would appear to be about that expected from the NCHS standards. As with the Bundi children (Malcolm, 1971), this has not come about by the maintenance of the deficit acquired in childhood. These girls initially gained height and then fell back at an early puberty, whereas the Bundi children lost relative height and then caught up later.

Another source of data to examine the capacity for catch-up growth is affluent children with stunting secondary to disease. This has the advantage that these groups are less likely to have had early developmental problems and malnourished parents. On the other hand, even with treatment, their residual disease, previous surgery, special diets and drugs may each restrict their recovery. There are many case reports of impressive catch-up that have been previously reviewed (Prader, Tanner & Harnack, 1963; Tanner, 1981; Prader, 1978). For example, two hypopituitary dwarfs were treated with growth hormone and androgen when aged 23 and 24 years: they subsequently gained 22 and 21 cm (Van der Werff Ten Bosch & Bot, 1986). Another patient who had constrictive pericarditis corrected surgically at the age of 18 subsequently gained 32 cm (Van den Brande, 1986). These reports serve to illustrate that catch-up is possible even at a chronological age where growth has normally ceased; where catch-up is not observed, it does not mean that it is not possible, only that we have not yet found the correct way of bringing it about. Most children with stunting secondary to renal or intestinal disease fail to catch up to their expected adult height.

Prader's group (Barr, Shmerling & Prader, 1972) studied growth in coeliac patients aged 9–15 months after institution of treatment. At this age, catch-up in height and bone age, which were similarly retarded, was complete; it took about two years. Children over 3 years of age, when diagnosed, although severely stunted, usually have heights that are in keeping with their bone ages and so complete catch-up growth should be achievable (Exner et al.,1978); however, in other centres, about half these children fail to have any significant catch-up (Fagundes-Neto, Stump & Wehba, 1981).

Older children (11 to 17 years) with Crohn's disease show a marked height spurt after surgery (Lipson et al., 1990). Follow-up of a different group of children with Crohn's disease into adult life showed that most attain normal adult height (Castile, Telander & Cooney, 1980; Markowitz et al., 1993). Permanent growth failure was much less common in subjects with stunting due to ulcerative colitis (10%) (Markowitz et al., 1993). This agrees with the findings of Kirschner (Kirschner, Voinchet & Bosenberg, 1978).

The complete reversibility of the clinical disease, without need for continued treatment, characterizes children with the trichuris dysentery syndrome. Cooper et al. (1990) gave such children anthelminthic treatment without any other change in their circumstances. The children had an enormous height spurt. This study is important because it shows that children living in poor conditions in countries where stunting is common can, even in later childhood, have a very substantial catch-up in height.

It is quite clear that very substantial catch-up in height to totally eradicate a deficit is possible long after the early stages of growth, defined by Karlberg, are complete. There have been no studies conducted where optimum nutrition (Golden, 1991) has been provided to chronically stunted children, so we just do not know what the potential for catch-up in most groups of stunted children. This, of course, begs the question of 'what is the height potential?'.

Environment versus genetic potential

By comparing elite peoples from around the world, Martorell has produced convincing evidence that the genetic component of adult height is, on average, trivial in comparison to the environmental influences. Nevertheless, the height potential of an individual is related to the height of the parents. Before addressing the question of how complete the catch-up growth of an individual is, we need to know that individual's potential. In impoverished communities, this can lead to a circular argument. The parents were short because of malnutrition and this, in part, determines the target that the child is aiming towards (his individual potential). If he reaches this target, has he had complete catch-up? If so, is the child then normal? These questions show two different concepts of catch-up. One is a catch-up to what is expected of the individual child and the other is catch-up to the NCHS standard: they may be very different.

The analysis of Martorell where he attempts to disaggregate the genetic and environmental 'components' of stunting is generally followed for many diseases with complex aetiologies that are supposed to have their origins in early life such as hypertension, obesity or diabetes. Twin studies are frequently used in these analyses, so that even such conditions as delinquency, criminality or homosexuality can have their genetic component assessed. Indeed, throughout the literature there is the consistent assumption that any defect is *either* due to an influence of the environment *or* to an 'abnormality' of genetic endowment, the so-called nature vs nurture divide, and that it is possible to apportion the relative amount of 'blame' between one or the other. There is an implicit corollary to this dichotomous way of thinking: the proportion of the condition that is determined by the genetic component is not amenable to treatment, except by manipulation of the genome itself, whereas the environmental component is readily amenable to intervention strategies. Such ideas underlie the analysis of height gain, catch-up and potential in an individual or population, as well as the wider ideas of the fetal origins of adult chronic disease. I reject both this dichotomous 'either–or' concept of disease causation and also the corollary that any genetic or otherwise 'programmed' component will inexorably lead to illness which cannot be avoided because it is 'in the genes'.

Consider phenylketonuria; here is a well-understood hereditary condition that is caused by a single gene defect, undoubtedly, generally thought of as a condition *that is entirely due to a genetic defect*. But what if the whole population, in which the aberrant gene occurred, subsisted upon a low protein, and hence low phenylalanine, diet; in this case the disease would not occur and those individuals carrying the gene would never be identified. Thus, whether the disease occurs or not is entirely dependent upon the protein intake of the population carrying the gene, in other words it is logically just as legitimate to say that *the disease is entirely environmental*. It is absurd to have one disease that can be considered to be entirely environmental or entirely genetic depending, equally entirely, upon which point we choose to start to think about the disease and which discipline we adhere to. The alternative model of disease is much more complex and difficult to analyse, it is a completely interactive one where neither the genetic component nor the environment are to 'blame' for any condition and that all disease is an interaction between the two components. In this model the 'measured' proportion of a condition that is due to genetic endowment is dependent upon the environment in which it is measured, and the proportion that is environmental depends upon the genetic make-up of the population under study. Thus, the genetic contribution to road accidents (an environmental event) might well depend upon the prevalence of genetically determined poor eyesight or psychosis within a population. Would the difference between mono- and dizygous twins, in some characteristics such as height, be the same if the twins was reared in separate homes in the western world and where one twin was reared in England and the other with the Bundi in the New Guinea highlands? Are the environmental and genetic components of heart disease among East Indians the same when measured in India and the UK (Bhatnagar *et al.*, 1995) or for diabetes among Pima indians who are living a western or traditional lifestyle? Such 'thought experiments' demonstrate that it is meaningless to try to separate a genetic and environmental *component* to such conditions as stunting when we should be directing our efforts towards understanding the nature of their interaction.

Most situations are very much more complex than phenylketonuria. This is illustrated in a study of variations of Apolipoprotein A1 in a baboon colony by Blangero *et al.*, 1990. Two major genetic loci were considered in relation to either a normal or an atherogenic diet. There was a diet × gene interaction at both loci so that, although most of the genotypes had a higher Apo A1 on the atherogenic diet, some of the genotypes had a lower level on this diet. In this analysis only two loci and two diets were considered, most real-life situations involve a complete intertwining of genetic and environmental influences. Our experiments and interpretations are usually, of necessity, simplistic. The human counterpart of the baboon experiment is illustrated by Kaprio *et al.* (1989) when they found that the effect of genetic variation of a cholesterol ester transfer protein (*Taq*I RFLP) upon apo-A1 serum levels was apparent in non-smokers but totally absent in smokers.

The classical experiments of cross-breeding shire horses with Shetland ponies (Walton & Hammond, 1938) dramatically demonstrate the effect both genetic endowment and intrauterine environment have upon the subsequent growth of animals. The adult progeny weighed half as much again when the Shetland was the sire rather than the dam. Similar experiments with embryo-transplanted sheep have confirmed these findings (Hammond, 1960). The human equivalent is the fate of monozygotic twins of different size; the smaller twin remains small and fails to catch up as an adult (Babson & Phillips, 1973).

The complex interaction of ram and ewe genetic influences and the plane of maternal nutrition has been examined by Hammond in sheep (Hammond, 1960). His data show that influences on growth which principally operate postnatally, such as litter size, dominate for the months after birth but then these influences wash out, so that at 8 months of age they are negligible. The intrauterine environment and genetic influences are clearly discernible in later life, as well as a large amount of 'unexplained' variation.

There are no indigenous wild pigs in New Zealand. Captain Cook introduced English domestic pigs in 1772, some of which escaped to become feral pigs. Today they are almost indistinguishable from wild boar in conformation and composition. The genetic differences are minor between these animals and their domestic cousins, but the phenotypic differences are enormous. Despite having to forage for food instead of having abundant food supplied, the feral pigs' bones are considerably longer (and thinner) than domestic pigs of the same carcase weight (Hammond, 1960).

It is not only the plane of nutrition that affects growth and conformation. Porcine litter mates that are reared at 35°C are much taller and thinner, at the same carcase weight, than pigs that are raised at 5°C (Mount, 1968).

Several years ago Stewart, Preece and Sheppard (1975) divided a colony of rats into two. One-half were given a restricted protein diet and the other half a control diet. The progeny of the restricted rats were smaller than the control rats and remained so over many generations; it was as if they were a different strain of rat all together. When the restricted rats were returned to the control diet, subsequent generations were larger, but it took three generations for them to attain the size of the control colony (Stewart *et al.*, 1980).

Inherited effects that gradually wash out over several generations have been seen in other functions apart from growth. Beach, Gershwin and Hurley (1982) gave mice a brief period of zinc deficiency *in utero* and then cross-fostered the pups at birth. They were immuno-deficient as adults: their progeny were also partially immuno-deficient; normality was virtually, but not quite, restored by the third generation.

The usual explanation for the shire horse/ Shetland phenomenon and that observed in Stewart's rats is that the mother's small size deprives the fetus of nutrients through a small placenta and reduced uterine blood flow.

Yet, the foals are quite normal except for their size, they do not demonstrate the usual stigmata of malnutrition. This explanation is neither sufficient for the size variation of the foals or for the 'inherited' immuno-incompetence of the mice whose grandparent was given a zinc-deficient diet. How could an insult to the grandmother during pregnancy affect the immuno-competence of the subject?

It is now becoming clear that the interaction between the genome and the environment is much more direct than the simple example of phenylketonuria used to reject the either-nature-or-nurture concept and propound the idea of all conditions being interactive. This is because the maternal (and paternal) environments affect the degree and timing of expression of genes in their offspring – an epigenetic phenomenon known as imprinting (Hall, 1991; Solter, 1988). The mechanism of imprinting is through methylation of the DNA bases (Surani *et al.*, 1990) which switch the genes off or on during early development. For example, although we inherit a gene from our mother and our father, the one expressed, and the timing and degree of its expression, is determined by epigenic modification during early development, including meiosis when the gametes are being formed. For example, only the gene for insulin-like growth factor II derived from our father is expressed, whereas only the maternal gene for the receptor of this growth factor is expressed (Haig & Graham, 1991).

Control of DNA base methylation and the substrates and coenzymes involved in methylation reactions warrant examination in relation to height 'potential'. The factors that affect base methylation include the supply of methionine, serine, glycine, folic acid (5'-methyl-tetrahydrofolate), cobalamin, pyridoxine, betaine and choline as well as several enzymes known to exhibit considerable variation in the population.

We are now beginning to appreciate the crucial importance of interaction between nutrition and genetic expression. For example, Cheviot sheep are a breed that does not have horns and has not had horns since the selective breeding that established the breed many generations ago; when they are put on a zinc-deficient diet, they start to grow horns (Mills *et al.*, 1967), despite the fact that zinc deficiency suppresses protein synthesis and growth generally. Although this breed of sheep does not normally have horns, the appropriate genes are present but have been permanently switched off. However, the environmental insult, zinc deficiently, directly causes the expression of the hidden genes to be unmasked. Presumably, this is either due to an error in the fidelity of replication or to the removal of a very specific transcription inhibition.

Although retinoic acid, cholecalciferol metabolites and the zinc finger proteins, which are all under nutritional influence, affect the expression of genes directly, these nutritional influences are not thought to switch genes on and off permanently in the way that imprinting through base methylation does.

Variation in imprinting, during very early embryonic development, by specific epigenomic modification, provides a satisfactory explanation for Hammond's and Stewart's experiments, as well as the studies on zinc deficiency. As oogenesis occurs in fetal life, the nutritional plane of the grandparents may indeed be as important as that of the mother. If the meiotic and early *in utero* environment changes the pattern of epigenic modification to alter the potential for future somatic development, over several generations, it would also offer a satisfactory explanation for the close association between height of a child and of its parents and yet allow for the both secular trend and the height of elite peoples around the world to be similar.

The phenomenon of the increasing secular trend in height clearly transcends the sequence of bases in the germline, which is relatively stable. The purely genetic hypothesis would require the base sequence to vary randomly so that the offspring are both smaller and larger than their parents; a secular trend could then emerge if the larger ones were selected by differential mortality or breeding success. There is no evidence that such a mechanism underlies secular trend in height. In contrast to this Darwinian explanation, children are consistently taller than their parents during the population increases in height. And yet, it is very difficult to envisage an environmental or nutritional mechanism for the smooth, gradual and small incremental changes in height from one generation to the next, without invoking a complex and subtle gene × nutrient interaction. It makes sense teleologically for the parents to be able to 'program' the child according to their own interactions with the environment; parenterally derived programming becomes a hindrance when the environment of the offspring is entirely different from that of the parent. Such phenomena may underlie the cardiovascular problems of Indian immigrants to the United Kingdom, or of the changes in height and early menarche of Indian adoptees to Sweden. Certainly, gross changes in placental blood flow, supply of bulk nutrients such as total protein to the fetus or uterine size are not satisfactory explanations for secular changes.

The different controls of growth at different periods of growth (Karlberg, 1989) may indeed mean that, if abnormal growth is programmed by very early developmental events to occur during a particular period, there may not be the capacity or stimulus to catch up when the child enters a further period. Martorell's data on the increments in Guatemalan children from 5 to 18 years of age (Martorell *et al.*, 1990) suggest that these subjects do not have the inherent 'drive' for increased growth; although it could equally be due to continued deficiency of a type II nutrient.

Nevertheless, the failure of catch-up of the Guatemalan children may well be unrelated to postnatal experience and be due to early developmental events, whereas children stunted postnatally may be able to undergo complete catch-up. This would agree with Hammond's findings in sheep (Hammond, 1960) and be a sufficient explanation for the complete catch-up by the African groups of stunted malnourished children. An immense amount of experimental work remains to be done to understand fully why some children fail to catch up whilst others undergo a complete and full recovery despite continuing seeming privation. For some groups to fully catch up, we may have to correct a defect over several generations. One can view a positive secular trend as a cross-generational catch-up in height. If we observe a malnourished child who is indeed short and yet surpasses his potential, defined by parental height, we should say that he has had a complete catch-up for himself as an individual (even though he fails to reach the NCHS standard), but not a complete catch-up as a pedigree.

It is clear that, in the third world, there is often an adverse prenatal development, and that many of the children that later present with severe stunting may indeed have a lower growth potential.

References

Alvear, J., Artaza, C., Vial, M., Guerrero, S. & Muzzo, S. (1986). Physical growth and bone age of survivors of protein energy malnutrition. *Archives of Disease in Childhood*, **61**, 257–62.

Babson, S.G. & Phillips, D.S. (1973). Growth and development of twins dissimilar in size at birth. *New England Journal of Medicine*, **289**, 937–40.

Baertl, J.M., Adrianzen, B. & Graham, G.G. (1976). Growth of previously well-nourished children in poor homes. *American Journal of Diseases of Childhood*, **130**, 33–6.

Barr, D.G.D., Shmerling, D.H. & Prader, A. (1972). Catch-up growth in malnutrition, studied in celiac disease after institution of gluten-free diet. *Pediatric Research*, **6**, 521–7.

Beach, R.S., Gershwin, M.E. & Hurley, L.S. (1982). Gestational zinc deprivation in mice: persistence of immunodeficiency for three generations. *Science*, **218**, 469–71.

Bhatnagar, D., Anand, I.S., Durrington, P.N., Patel, D.J., Wander, G.S., Mackness, M.I., Creed, F., Tomenson, B., Chandrashekhar, Y., Winterbotham, M., Britt, R.P. & Keil, J.E. (1995). Coronary risk factors in people from the Indian subcontinent living in West London and their siblings in India. *Lancet*, **345**, 405–9.

Binkin, N.J., Yip, R., Fleshood, L. & Trowbridge, F.L. (1988). Birth weight and childhood growth. *Pediatrics*, **82**, 828–34.

Blangero, J., MacCluer, J.W., Kammerer, C.M., Mott, G.E., Dyer, T.D. & McGill, H. C. (1990). Genetic analysis of apolipoprotein A-1 in two dietary environments. *American Journal of Human Genetics*, **47**, 414–28.

Bowie, M.D., Moodie, A.D., Mann, M.D. & Hansen, J.D. L. (1980). A prospective 15-year follow-up study of kwashiorkor patients. Part I. Physical growth and development. *South African Medical Journal*, **58**, 671–6.

Castile, R.G., Telander, R.L. & Cooney, D.R. (1980). Crohn's disease in children: assessment of the progression of disease, growth and prognosis. *Journal of Pediatric Surgery*, **15**, 462–9.

Cooper, E.S., Bundy, D.A.P., MacDonald, T.T. & Golden, M.H. (1990). Growth suppression in the *trichuris* dysentery syndrome. *European Journal of Clinical Nutrition*, **44**, 285–91.

Dreizen, S., Spirakis, C.N. & Stone, R.E. (1967). A comparison of skeletal growth and maturation in undernourished and well nourished girls before and after menarche. *Journal of Pediatrics*, **70**, 256–63.

Exner, G.U., Sacher, M., Shmerling, D.H. & Prader, A. (1978). Growth retardation and bone mineral status in children with coeliac disease recognized after the age of 3 years. *Helvetica Paediatrica Acta*, **33**, 497–507.

Fagundes-Neto, U., Stump, M.V. & Wehba, J. (1981). Catch-up growth after the introduction of a gluten-free diet in children with celiac disease. *Arquivos Gastroenterology*, **18**, 30–4.

Goel, K.M., Thomson, R.B., Sweet, E.M. & Halliday, S. (1981). Growth of immigrant children in the centre of Glasgow. *Scottish Medical Journal*, **26**, 340–5.

Golden, M.H. (1988). The role of individual nutrient deficiencies in growth retardation of children as exemplified by zinc and protein. In *Linear Growth Retardation in Less Developed Countries*, ed. J.C. Waterlow, pp. 143–63. New York: Raven Press.

Golden, M.H. (1991). The nature of nutritional deficiency in relation to growth failure and poverty. *Acta Paediatrica Scandinavica*, **374**, 95–110.

Graham, G.G., Adrianzen, B., Rabold, J. & Mellits, E.D. (1982). Later growth of malnourished infants and children. Comparison with 'healthy' siblings and parents. *American Journal of Diseases of Childhood*, **136**, 348–52.

Graham, G.G. & Adrianzen, B. (1971). Growth, inheritance and environment. *Pediatric Research*, **5**, 691–7.

Graham, G.G. & Adrianzen, T.B. (1972). Late 'catch-up' growth after severe infantile malnutrition. *Johns Hopkins Medical Journal*, **131**, 204–11.

Haig, D. & Graham, C. (1991). Genomic imprinting and the strange case of the insulin-like growth factor II receptor. *Cell*, **64**, 1045–6.

Hall, J.G. (1991). Genomic imprinting. *Current Opinion in Genetics and Development*, **1**, 34–9.

Hambidge, K.M. (1986). Zinc deficiency in the weanling, how important. *Acta Paediatrica Scandinavica*, **323**, 52–8.

Hammond, J. (1960). *Farm Animals: Their Breeding, Growth and Inheritance*, 3rd edn. London: Edward Arnold.

Kaprio, J., Ferrell, R.E., Kotte, B.A. & Sing, C.F. (1989). Smoking and reverse cholesterol transport: evidence for gene–environment interaction. *Clinical Genetics*, **36**, 266–8.

Karlberg, J., Jalil, F. & Lindblad, B.S. (1988). Longitudinal analysis of infantile growth in an urban area of Lahore, Pakistan. *Acta Paediatrica Scandinavica*, **77**, 392–401.

106   *M.H.N. Golden*

Karlberg, J. (1989). On the construction of the infancy-childhood-puberty growth standard. *Acta Paediatrica Scandinavica*, **356**, 26–37.
Keet, M.P., Moodie, A.D., Wittmann, W. & Hansen, J.D. L. (1971). Kwashiorkor: a prospective ten-year follow-up study. *South African Medical Journal*, **45**, 1427–49.
King, J.M. & Taitz, L.S. (1985). Catch-up growth following abuse. *Archives of Disease in Childhood*, **60**, 1152–4.
Kirschner, B.S., Voinchet, O. & Bosenberg, I.H. (1978). Growth retardation in inflammatory bowel disease. *Gastroenterology*, **75**, 504–11.
Kulin, H.E., Bwibo, N., Mutie, D. & Santner, S.J. (1982). The effect of chronic childhood malnutrition on pubertal growth and development. *American Journal of Clinical Nutrition*, **36**, 527–36. (Abstract)
Lien, N.M., Meyer, K.K. & Winick, M. (1977). Early malnutrition and 'late' adoption: a study of their effects on the development of Korean orphans adopted into American families. *American Journal of Clinical Nutrition*, **30**, 1734–9.
Lipson, A.B., Savage, M.O., Davies, P.S., Bassett, K., Shand, W.S. & Walker Smith, J.A. (1990). Acceleration of linear growth following intestinal resection for Crohn disease. *European Journal of Pediatrics*, **149**, 687–90.
Little, M.A., Galvin, K. & Mugambi, M. (1983). Cross-sectional growth of nomadic Turkana pastoralists. *Human Biology*, **55**, 811–30.
Malcolm, L.A. (1971). *Growth and Development in New Guinea: A Study of the Bundi People of the Madang District*. Madang: Institute of Biology.
Markowitz, J., Grancher, K., Rosa, J., Aiges, H. & Daum, F. (1993). Growth failure in pediatric inflammatory bowel disease. *Journal of Pediatric Gastroenterology and Nutrition*, **16**, 373–80.
Martorell, R., Yarbrough, C., Klein, R.E. & Lechtig, A. (1979). Malnutrition, body size, and skeletal maturation: interrelationships and implications for catch-up growth. *Human Biology*, **51**, 371–89.
Martorell, R., Rivera, J. & Kaplowitz, H. (1990). Consequences of stunting in early childhood for adult body size in rural Guatemala. *Annales Nestle*, **48**, 85–92.
Martorell, R., Rivera, J., Kaplowitz, H. & Pollitt, E. (1992). Long-term consequences of growth retardation during early childhood. In *Human Growth: Basic and Clinical Aspects*, ed. M. Hernandez & J. Argente, pp. 143–9, Amsterdam: Elsevier Science Publishers.
Mills, C.F., Delgarno, A.C., Williams, R.B. & Quarterman, J. (1967). Zinc deficiency and zinc requirement of calves and lambs. *British Journal of Nutrition*, **21**, 751–68.
Mills, J.L., Shiono, P.H., Shapiro, L.R., Crawford, P.B. & Rhoads, G.G. (1986). Early growth predicts timing of puberty in boys: results of a 14-year nutrition and growth study. *Journal of Pediatrics*, **109**, 543–7.
Mjones, S. (1987). Growth in Turkish children in Stockholm. *Annals of Human Biology*, **14**, 337–47.
Mount, L.E. (1968). *Climatic Physiology of the Pig*, London: Edward Arnold.
Pelletier, D.L., Low, J.W. & Msukwa, L.A.H. (1991). Malawi maternal and child nutrition study: study design and anthropometric characteristics of children and adults. *American Journal of Human Biology*, **3**, 347–61.
Prader, A., Tanner, J.M. & Harnack, G.A.V. (1963). Catch-up growth following illness or starvation. An example of developmental canalization in man.

Journal of Pediatrics, **62**, 646–59.

Prader, A. (1978). Catch-up growth. *Postgraduate Medical Journal*, **54**, Suppl 1, 133–46.

Proos, L.A., Hofvander, Y. & Tuvemo, T. (1991*a*). Menarcheal age and growth pattern of Indian girls adopted in Sweden. II. Catch-up growth and final height. *Indian Journal of Pediatrics*, **58**, 105–14.

Proos, L.A., Hofvander, Y. & Tuvemo, T. (1991*b*). Menarcheal age and growth pattern of Indian girls adopted in Sweden. I. Menarcheal age. *Acta Paediatrica Scandinavica*, **80**, 852–8.

Satyanarayana, K., Nadamuni Naidu, A. & Narasinga Rao, B.S. (1980). Adolescent growth spurt among rural Indian boys in relation to their nutritional status in early childhood. *Annals of Human Biology*, **7**, 359–65.

Satyanarayana, K., Nadamuni Naidu, A., Swaminathan, M.C. & Narasinga Rao, B.S. (1981). Effect of nutritional deprivation in early childhood on later growth: community study without intervention. *American Journal of Clinical Nutrition*, **34**, 1636–7.

Satyanarayana, K., Prasanna-Krishna, T. & Narasinga-Rao, B.S. (1986). Effect of early childhood undernutrition and child labour on growth and adult nutritional status of rural Indian boys around Hyderabad. *Human Nutrition: Clinical Nutrition*, **40**, 131–9.

Satyanarayana, K., Radhaiah, G., Murali Mohan, R., Thimmayamma, B.V., Pralhad Rao, N. & Narasinga Rao, B.S. (1989). The adolescent growth spurt in height among rural Indian boys in relation to childhood nutritional background: an 18 year longitudinal study. *Annals of Human Biology*, **16**, 289–300.

Schumacher, L.B., Pawson, I.G. & Kretchmer, N. (1987). Growth of immigrant children in the newcomer schools of San Francisco. *Pediatrics*, **80**, 861–8.

Solter, D. (1988). Differential imprinting and expression of maternal and paternal genes. *Annual Review of Genetics*, **22**, 127–46.

Steckel, R.H. (1987). Growth depression and recovery: the remarkable case of American slaves. *Annals of Human Biology*, **14**, 111–32.

Stewart, R.J., Preece, R. & Sheppard, H. (1975). Twelve generations of marginal protein deficiency. *British Journal of Nutrition*, **33**, 233–53.

Stewart, R.J., Shephard, H., Preece, R. & Waterlow, J.C. (1980). The effect of rehabilitation at different stages of development of rats marginally malnourished for ten to twelve generations. *British Journal of Nutrition*, **43**, 403–12.

Surani, M.A., Allen, N.D., Barton, S.C., Fundele, R., Howlett, S.K., Norris, M.L. & Reik, W. (1990). Developmental consequences of imprinting of parental chromosomes by DNA methylation. *Proceedings of the Royal Society (Series Biol)*, **326**, 313–27.

Tanner, J.M. (1981). Catch-up growth in man. *British Medical Bulletin*, **37**, 233–8.

Van den Brande, J.L. (1986). Catch up growth: possible mechanisms. *Acta Endocrinologica Supplement (Copenhagen)*, **279**, 13–23.

Van der Werff Ten Bosch, J.J. & Bot, A. (1986). Growth Hormone and androgen effects in the third decade. *Acta Endocrinologica Supplement (Copenhagen)*, **279**, 29–34.

Walravens, P.A., Krebs, N.F. & Hambidge, K.M. (1983). Linear growth of low income preschool children receiving a zinc supplement. *American Journal of*

Clinical Nutrition, **38**, 195–201.

Walravens, P.A., Hambidge, K.M. & Koepfer, D.M. (1989). Zinc supplementation in infants with a nutritional pattern of failure to thrive: a double-blind, controlled study. *Pediatrics*, **83**, 532–8.

Walravens, P.A. & Hambidge, K.M. (1976). Growth of infants fed a zinc supplemented formula. *American Journal of Clinical Nutrition*, **29**, 1114–21.

Walton, A. & Hammond, J. (1938). The maternal effects on growth and conformation in Shire horse–Shetland pony crosses. *Proceedings of the Royal Society (Series Biol)*, **125**, 311–35.

Winick, M., Meyer, K.K. & Harris, R.C. (1975). Malnutrition and environmental enrichment by early adoption. *Science*, **190**, 1173–5.

7 Influence of under-nutrition in early life on growth, body composition and metabolic competence

S.A. WOOTTON AND A.A. JACKSON

The normal growth and development of the conceptus is programmed as an intrinsic feature of the genome, with an orderly pattern to the sequence and timing of events being a fundamental aspect of the maturational process. This pattern of change and development takes place at each level of organization, from the molecular to the subcellular, the cellular to the level of the tissue or organ, to be ultimately coordinated at the level of the whole body. Although the blueprint for the process is maintained within the genome, this can only be translated into reality if the nutrients which provide the energy, substrate and cofactors are present in adequate amounts, and the cell signalling mechanisms are intact.

The general concept that diet during early life might influence the rate of growth and development at a later stage is not new (Widdowson & McCance, 1963). Nor is the idea that there are periods of time during which one or other function is particularly sensitive to an adverse insult and therefore the consequence of the insult might have lasting effect. It is known that the susceptibility to a potentially damaging insult varies with a number of factors, in particular the nature and severity of the stress, the timing and duration of its action. As organogenesis takes place during the embryonic period, the rate of cell multiplication and organization takes place at such a pace that the organism is especially sensitive, and the cost of failure is either death or a severe on structural development, manifest as teratological change. At later stages, once the gross structure of the organ systems has been established, the manifestation of the insult may be more subtle. With time, the plasticity of the events becomes less, which means that each system tends to be more sensitive to external perturbations the earlier the perturbation operates in the development of that particular system.

Classically, the approach of teratology has been to explore the effect of severe perturbations upon the system and it is readily demonstrable that the detailed effect as a result of an insult is determined by the nature of the insult, but also by the timing, the severity and the duration for which the insult operates. In grossest terms, this is seen as a limitation in the failure of an organ to form to grow adequately in size of shape. More subtle changes which might be attributed to a reduction in the metabolic capacity of a system to carry out a particular function may be more difficult to identify, simply because, as a matter of course, it is normal for excess functional capacity (functional redundancy) as an intrinsic feature of most systems. Under these circumstances the limit of the capacity is not evident under normal operating conditions, and the limitation can only be exposed by an appropriate stress test which assesses the maximal capacity of the system to respond.

The characteristics of normal growth and development is an increase in the size and the complexity of function of the tissues and organs of the body. The increase in size represents a combination of the increase in the number of cells and the increase in the size of each individual cell. In its simplest form, this is the basis of the representation of three phases of growth (Winick & Noble, 1966), pure hyperplasia, mixed hypertrophy and hyperplasia and pure hypertrophy. Winick identified that, for any tissue, the timing of the three phases is fixed, but that the timing might vary from one tissue to another and that any particular tissue is most sensitive to an irreversible insult which acts during the period of time of most rapid cell division, because of an ultimate limitation on the functional mass of the tissue. The theory considered that, although damage might occur as a consequence of an insult acting during a period of increase in cell size, the effect of adverse influences acting at this time are more likely to be reversible when the insult is removed.

Dobbing has suggested that this useful concept is too restrictive when applied uncritically (Dobbing, 1990). The theory implies that all limitation of function is directly related to a restriction in size, mass or form: an anatomical of morphological model. As a matter of simple observation, cell multiplication takes place over a much longer period than allowed for by the theory, and for long periods of time the organism appears to be less sensitive to stress than would be expected. Dobbing preferred an interpretation which allowed for a dependence upon the process associated with the maturation of organ function, a physiological model. As others have considered the issue in greater detail, more subtle mechanisms through which an insult might act have been explored. For example, Lucas (1991) has identified three potential cellular mechanisms which include the older ideas of reduction in cell numbers and permanent alteration of gene expression, but also the idea of clonal selection of population of cells an early stage (akin to the ideas of clonal selection in immunology). The extent to which clonal selection, or other mechanisms which impact upon cell switching might operate individually or collectively, remains to be explored in detail. What is clear is that the problem can be characterized and investigated at different levels of organization: anatomical, physiological, biochemical, immunological. It is of interest to explore the nature of the involvement of nutrition, a metabolic consideration of integration at the level of the whole body. How can a metabolic model be created which allows for a more complex interplay of factors in time, taking account of the varied pattern of maturation of metabolic function, the acquisition of metabolic memory and the determinants of expression of DNA within the context of historical experience?

The essence of nutrition is the idea that nutrient balance is achieved within a dynamic framework. There is a constant internal exchange of material, which interacts with an external exchange, the diet, to ensure that the tissues are constantly nourished. That is the tissues are provided with the energy, substrate and cofactors which enable them to maintain function. This function represents a metabolic demand and it is the metabolic demand which has to be satisfied by the dietary and endogenous provision of substrate and cofactors. Hence the programming of metabolic memory has to include a cellular recognition of what is perceived by the organism as an appropriate balance of inputs and outflows at a critical point in time. Conceptually, this is similar to a general idea which in the past has been described, non-specifically, as the plane of nutrition. Any model of growth has to take into account the process of development in form and function, but also it has to recognize this as one part of a continuous process of metabolic change which in addition includes a consideration of the processes of ageing, that is the loss of function or degenerative disease (Barker, 1992).

Any general model of growth represents an increase in functional capacity. As the functional capacity defines the upper limit of the metabolic capability, the maximal capability of the tissue or organ to carry out a function, it can only be measured or determined under conditions of metabolic load. The achieved capacity will relate to the number of cells and the sum of their size, but the evidence shows that it also relates to the metabolic experience of those cells. Within this construct growth and development are seen as an increase in, and maturation of metabolic functional capacity with time, up to an adult capacity. Functional capacity increases during growth as the number and size of cells increase. Early adulthood is characterized by the maximum metabolic capability (capacity) the individual is likely to achieve for many metabolic functions. During normal life, the extent to which the maximal capacity is called upon varies, but for most situations or functions there is a significant reserve capacity which is only called into play under unusual situations of stress, or unusually when the metabolic demand is significantly increased.

Ageing represents a progressive loss of this functional capacity, not necessarily in terms of mass, but more likely in terms of metabolic function at a cellular level (concentration per unit cell decreases). Given the excess capacity normally developed, a significant reserve capacity may be maintained for some time, but as the capacity is progressively lost the margin of safety decreases, hence, the ability to cope with environmental perturbations becomes more limited, and the risk of the capacity being exceeded increases. Everyday function may be maintained to a very late stage under these circumstances, but a limitation of function may well be identified with a stress test designed to assess the magnitude of the reserve capacity. It is not clear whether the rate of loss is constant or some function of the total capacity, but for a given rate of loss the individual with a lower peak function will be at greater risk and reach the limit of ability to cope at an earlier time than one with higher peak capability. Degenerative disease is perceived as a progressive loss of functional capacity relative to the metabolic demand of the system as a whole (Figure 7.1). That is, disease supervenes when the functional capacity is insufficient to cope with the metabolic demands being placed on the organism as a whole: the more limited the final adult capacity, the earlier the age at which the critical threshold is reached, below which the system cannot cope.

Restricted nutrition during early growth

Restricted nutrition at any stage of life produces a series of changes in the organism which together facilitate survival for as long as possible. There is a reduction in the energy consumption of the organism, a process of metabolic adaptation at the organ and cellular level, through a reduction in the rate at which metabolic processes turn over, which may constrain the net deposition of new tissue. If sufficiently gross, the limitation of deposition of new tissue is identifiable as impaired cellular growth and development. The integration and harmonization of growth in time requires that earlier events are successfully achieved if later events are to take place in an orderly and successful way. Therefore, the functional consequences of restricted nutrition in early life may be most readily identified as impaired growth and development. More subtle effects, which constrain the function of the systems which integrate metabolism, acting during the critical period of development when a function is particularly sensitive to an adverse insult might have lasting effect on metabolic integration and persist through to adult life (Widdowson & McCance, 1963; MaCance & Widdowson, 1974). As this is part of a continuous

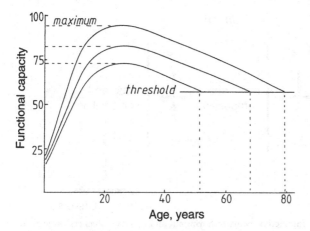

Figure 7.1. Schematic representation of the changes in functional capacity with age. Degenerative disease supervenes when the functional capacity is insufficient to cope with the metabolic demands of the body.

process of metabolic change, it will in turn dictate both the metabolic demand and the handling of ingested energy and nutrients. It is proposed that even a modest limitation of nutrient availability in early life, even within the ranges of normal energy and nutrient intake, will influence the achieved mass and composition of the infant and that this in turn will compromise the maximal functional capacity that an individual may achieve as an adult.

Growth and development in early life are modulated by the interaction between the hormonal milieu and the supply of energy and nutrients, which together determine the extent to which genetic endowment is expressed (Figure 7.2). Achieved body size and composition of the newborn infant, in turn, influence the ability of the child to grow and develop in response to the dietary intake and hormonal influences. Differences in metabolic competence in adult life arise from differences in the mass and composition of the fetus which persist through into adult life. Thereafter, the extent to which the dietary intake matches the metabolic competence of the individual to handle substrate and cofactors influences the risk of disease.

Hormonal regulation of growth

At this stage it is worth exploring in more detail the interrelationships between growth, diet and hormonal regulation. The growth process is under the complex control of the endocrine system. Not just hormones such as those which may reflect short-term changes in the metabolic pattern (such as insulin) or those which may influence the rate of growth (such as

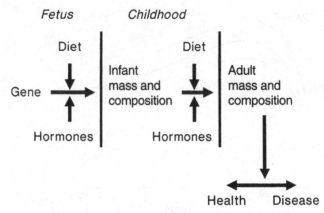

Figure 7.2. Interaction between hormonal and dietary influences on growth and development throughout early life.

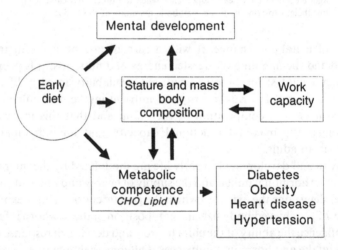

Figure 7.3. Infancy–childhood–puberty(IPC) growth model for boys showing the key hormones involved in each component (Karlberg, 1987). Note the inclusion of fetal growth within the Infancy component.

growth hormone, and sex steroids in puberty), but also hormone-binding proteins, local and circulating growth factors (Somatomedin and IGF-1) and their binding proteins as well as the maturity and number of hormone and growth factor receptors on the target cells (Cianfaranai & Holly, 1989; Waters *et al.*, 1990).

It is generally agreed that there are at least three distinct endocrine phases of linear growth. Karlberg (1987) breaks it down into three additive and superimposed components of linear growth from birth to maturity: infancy (this includes fetal growth), childhood and puberty, each regulated by different growth-promoting systems (Figure 7.3). How fetal growth is regulated is not precisely defined. Whilst the size of the uterus and supply of oxygen supply are clearly important, fetal growth is believed to be nutritionally driven by the supply of substrate and cofactors in conjunction with insulin and growth hormone (GH) as well as IGF (Gluckman, 1989; Milner & Hill, 1987; Hill, 1989). Fetal linear growth appears to be independent of GH, possibly as a result of immature GH-specific receptors.

It is more generally accepted that GH is responsible for growth during the childhood phase provided that thyroid hormone secretion is normal (Karlberg & Albertsson-Wickland, 1988). The exact age that GH begins to control linear growth is uncertain. The majority of children with GH deficiency grow normally for the first six months of life, but not thereafter. A postnatal continuation of the nutritionally driven fetal growth in conjunction with the GH-dependent phase of childhood growth character-izes growth in the first year of life. Growth during adolescence is related to both GH and sex steroids, testosterone and oestrogen, and the adolescent growth spurt is the product of the two superimposed phases of childhood and pubertal growth.

Karlberg proposes that any delay in the transition from infant to childhood growth will have profound effects on the attained height in subsequent years. This delay, or growth faltering, is seen more typically in developing countries and reflects socioeconomic standards, but may also be observed much closer to home. Identifying the causal factors is more difficult, but the importance of maintaining optimal infant growth has been highlighted in studies on normal Swedish infants. Karlberg and co-workers have shown that the rate of growth during infancy prior to the onset of childhood growth is negatively related to the age at onset of the childhood phase (Karlberg, Hagglund & Stromquist, 1991). The rate of infant growth must result from either a lower capacity to grow arising from uterine programming in terms of attained size (small babies grow more slowly), an inadequate energy and nutrient intake with respect to demand, altered endocrine regulation or some combination of all three possible factors.

The pattern of endocrine regulation, both in terms of timing and interaction, determines growth and development, and even modest changes in nutrition can result in profound metabolic adjustments.

Functional consequences of impairments in growth

Interest in the functional consequences of impairments in growth and development associated with malnutrition in early life are now the subject

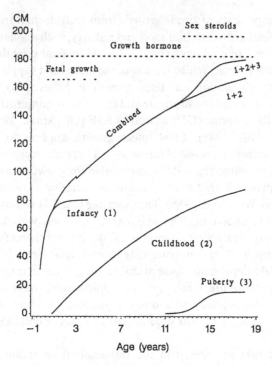

Figure 7.4. Schematic summary of the interaction between development in early life, physiological and metabolic competence and disease.

of considerable investigation. This schema is offered as a summary of our current way of looking at this problem (Figure 7.4).

Malnutrition in early life has been shown markedly to influence both cognitive function and psychomotor development (Grantham-McGregor & Walker, 1991). A causal relationship between severe malnutrition and poor mental development cannot be established unequivocally, but it is probable that a causal relationship exists most probably mediated by changes to the CNS or through changes in the child's exploration of their environment.

Impairment in growth and development are identified most readily as limitations in stature and muscularity. This, in turn, has been shown to influence the work capacity of the individual – primarily through a reduction in both strength and work output over time (Spurr, 1988). There is increasing evidence that the composition of skeletal muscle may also be modulated by malnourishment in early life with an increase in the proportion of fast glycolytic fibres (Type IIa) in preference to slow oxidative fibres (Type I) (Heymsfield *et al.*, 1982). Both the amount and fibre composition of skeletal muscle would markedly influence the functional and metabolic capacity of the individual throughout childhood and adult life. This, in turn, would influence the type and amount of habitual physical activity which would, of itself, also modulate body composition.

Attention is now being directed towards the extent to which differences in metabolic competence in terms of energy and substrate metabolism and physiological function may be explained through differences in mass and composition, particularly in terms of organ size and the relative proportions of skeletal muscle to viscera. Differences in metabolic competence may explain the observed differences in the way in which different individuals handle and dispose of the same dietary intake and the development of chronic disease such as non-insulin-dependent diabetes mellitus (NIDDM). For example, differences in the extent to which an oral load of carbohydrate may be oxidized or stored may be dependent on the relative proportions of lean body mass to adipose tissue, skeletal muscle to viscera or even the relative proportions of slow to fast skeletal muscle fibres.

Food intake and energy metabolism in relation to growth

At Southampton the interrelationships between body size and composition, energy and nitrogen metabolism and endocrine regulation during growth are being explored. In particular, the determinants of the functional capacity and metabolic competence of the individual and how these may be related to the perinatal experience by studying the metabolic consequences of poor growth and development in normal healthy children and chronically energy deficient adults are under scrutiny.

Size, in terms of stature, mass and composition all markedly influence energy metabolism. Long ago, Widdowson pointed out the twofold range in energy expenditure as basal metabolic rate (BMR) when expressed per unit body weight, in a group of normal children (Widdowson, 1962). One factor which may explain part of this variance is the apparent effect of height on BMR per unit body mass or lean tissue.

In an early study of short, so-called normal, children drawn from the normal population of Southampton, those with a Z score for height of −2, the BMR per unit lean body mass was about 30% greater than in control children at the 50th centile for height (Walker, Powell & Grantham-McGregor, 1990). Energy expenditure was inversely related to relative height for age (Z score). The interpretation of these data was that, relative to their size, the metabolic demands of short children were greater than those of their taller peers. It was suggested in subsequent correspondence about the paper that the difference in BMR could be eliminated by covariance analysis. Whilst this may make it possible to examine whether BMR was determined by any factor other than size, such as age or maturity, it may well obscure functional differences of physiological interest (Waterlow, 1992).

These observations of greater metabolic demands of short, otherwise healthy, children may explain the findings that stunted children in Jamaica had a 33% greater energy intake per kg than children of normal height in the same environment (Walker et al., 1990). Taken together, these two studies raise important questions as to the cause and effect relationships between food intake and poor growth: are short children short because their metabolic demands for energy and nutrients are raised, or are they short as an appropriate response to a limited availability of food?

These observations have been extended by studying growth and body composition, metabolic demand and dietary intake in short children and in short children receiving biosynthetic growth hormone compared to age and gender matched children of normal stature (Smith, 1994). When corrected for differences in lean body mass, both short stature groups exhibited greater energy intakes and BMR values than the control children (Table 7.1). Despite being more like the normal stature children in terms of height and weight, the short treated children had energy intake and BMR values more comparable to those of the short untreated children. It would appear that the metabolic demands of short children are greater than their peers of normal height and that this increased energy requirement is constitutive and remains with the short child despite the rapid growth engendered by growth hormone treatment.

Do children of short stature eat differently in response to these raised demands? Further analysis of the records of food intake of short children, short children receiving growth hormone and in children of normal height was conducted to characterize both the pattern of food intake as well as the nutrient intake. Although the relative proportions of carbohydrate, protein and fat in the diet were similar in all three groups as was the general nutrient density of the diet (Table 7.2), more subtle differences in the choice of foods consumed was observed. The short stature children consumed more confectionery, sugar, crisps and chips than the children of normal stature, either expressed as the relative contribution to total energy intake or corrected for differences in body weight. Following treatment with growth hormone, the short treated children increased their energy intake still further by consuming more cakes, biscuits, sugar, jam, confectionery, crisps and chips. Both short stature groups consumed fewer vegetables and less fruit than children of normal stature.

In this study, it would appear that short stature children have a preference for foods rich in carbohydrate, particularly sweet foods,

Table 7.1. *Subject details, reported energy intake and basal metabolic rate of normal healthy children of short stature (short untreated), children of short stature after 6 months treatment with biosynthetic growth hormone (short treated) and normal healthy children of normal stature (control)*

	Short untreated	Short treated	Control
Age (y)	9.2	9.4	9.7
Height (cm)	117*	124*	137
Height SDS	−2.5*	−1.5*	0.2
Weight (kg)	21.4*	24.4*	31.3
% Body fat	16	12*	20
Lean body mass (kg)	17.8*	21.3*	24.6
Energy intake			
(KJ/d)	6228*	8183	7719
(KJ/kgLBM/d)	353	387*	319
BMR			
(KJ/d)	4077*	4932	5192
(KJ/kglbm/d)	232	233*	212

*Significantly different from control $P < 0.05$.

compared to children of normal stature. Whether this pattern of food intake results from an altered innate pattern of substrate utilization associated with the increased metabolic demand or that shortness results from this pattern of food intake, or whether it simply reflects an altered provision of food by the carers of short stature children cannot be resolved from this study. Further studies are needed to examine more closely the partitioning of energy and nutrient intake in relation to body composition in children as they grow.

Chronic energy deficiency studies

Small children go on to become small adults. To examine the consequences of poor growth and development in adults, we can study individuals who have attained adulthood in the face of chronic energy deficiency. In a series of studies, Soares and co-workers have compared the energy and protein metabolism of chronically energy deficient (CED) unskilled manual labourers (CED is defined by low body mass index and socioeconomic group), with that of thin well-nourished controls and well-nourished

Table 7.2. *Contribution of macronutrients and major food groups to energy intake of normal healthy children of short stature (short untreated), children of short stature after 6 months treatment with biosynthetic growth hormone (short treated) and normal healthy children of normal stature (control)*

	Short untreated	Short treated	Control
% Energy from:			
Fat	36	34	35
Carbohydrate	53	55	53
Protein	11	11	12
% Energy from:			
Cakes and biscuits	12	17	12
Confectionery and sugars	20	21*	13
Crisps and chips	13*	12*	13
Vegetables and fruit	5*	5*	10
Meat and meat products	9	9	11
Milk and milk products		8	11

* Significantly different from control $P < 0.05$.

controls of normal body weight in India. The CED subjects were shorter, lighter and had a lower lean body mass, than either the thin or normal weight controls. Estimates of skeletal muscle mass based on urinary creatinine excretion showed that within the lean tissue, the chronically undernourished group had a higher proportion of non-muscle (viscera) to skeletal muscle mass (Soares & Shetty, 1991).

These differences in the composition of the lean tissue, with reductions in the metabolically less active skeletal muscle whilst the metabolically more active viscera was maintained, may explain the observed differences in energy and nitrogen metabolism between CED subjects and normal weight controls (Table 7.3). These differences include a higher BMR when expressed per kg body or lean body mass and a greater dependence on carbohydrate as a fuel in the fasted state. Studies of protein turnover using ^{15}N-glycine went on to reveal lower rates of protein synthesis and breakdown when expressed in absolute terms in undernourished subjects than normally nourished subjects on comparable dietary intakes (Soares *et al.*, 1994). As nitrogen flux (Q) was determined from ^{15}N abundance in two end products, it was possible to differentiate nitrogen metabolism through non-muscle and muscle tissues: urea (Qu) reflecting nitrogen flux through

Table 7.3. *Body composition, basal metabolic rate and respiratory quotient of well-nourished controls of normal weight (control), thin well-nourished controls (Thin) and chronically undernourished manual labourers (CED)*

	Control	Thin	CED
Height (m)	1.75	1.76	1.62* #
Weight (kg)	67.5	51.7*	43.6* #
LBM (kg)	53.8	44.6*	38.8* #
Muscle mass (kg)	31.7	22.2*	18.9* #
Non-muscle mass (kg)	22.2	22.4	19.9
Non-muscle:muscle	0.70	1.02*	1.08* #
BMR			
(KJ/kgBWt/d)	95.2	104.0	116.2* #
(KJ/kgLBM/d)	119.1	120.5	130.6* #
RQ	0.79	0.83	0.94*

Values from Soares & Shetty, 1991. * Significantly different from control $P < 0.05$; # significantly different from Thin $P < 0.05$.

organ tissues (mainly liver and gut) and ammonia (Qa) reflecting nitrogen flux through skeletal muscle. A higher Qu/Qa ratio was observed in the undernourished subjects and was significantly correlated with the ratio of non-muscle to muscle mass. Taken together, these observations illustrate the major role that size and composition may play in determining the metabolic characteristics of the body.

Conclusions

During early life, the available energy and nutrients interact with the hormonal milieu to determine pattern of growth, body composition and metabolic competence of the child and that these differences persist into adult life.

More research is required to determine how the mass and composition of the body determine the metabolic competence of the individual and the consequences of the more subtle variations of nutrient intake, energy and substrate metabolism observed in the population at large.

References

Barker, D.J.P. (1992). *Fetal and Infant Origins of Adult Disease*. London: British Medical Journal Publishing Group.

Cianfaranai, S. & Holly, J.M.P. (1989). Somatomedin-binding proteins: what role do they play in the growth process? *European Journal of Paediatrics*, **149**, 76–9.
Dobbing, J. (1990). Early nutrition and later achievement. *Proceedings of the Nutrition Society*, **49**, 103–18.
Gluckman, P.D. (1989). Fetal growth: an endocrine perspective. *Acta Paediatrica Scandinavica*. Suppl **349**, 21–5.
Grantham-McGregor, S. & Walker, S.P. (1991). Early diet and mental development. In: *Early Diet, Later Consequences*, ed. D. Conning. *British Nutrition Foundation Nutrition Bulletin*, **16**, Suppl 1, pp. 9–23.
Heymsfield, S.B., Stevens, V., Noel, R., McManus, C., Smith, J. & Dixon, D. (1982). Biochemical composition of muscle in normal and semistarved human subjects: relevance to anthropometric measurements. *American Journal of Clinical Nutrition*, **36**, 131–42.
Hill, D.J. (1989). Cell multiplication and differentiation. *Acta Paediatrica Scandinavica*, **349**, 13–20.
Karlberg, J. (1987). On the modelling of human growth. *Statistics in Medicine*, **6**, 185–92.
Karlberg, J. & Albertsson-Wickland, K. (1988). Infancy growth pattern related to growth hormone deficiency. *Acta Paediatrica Scandinavica*, **77**, 385–91.
Karlberg, J., Hagglund, B. & Stromquist, B. (1991). Immobilization related to early linear growth. *Acta Paediatrica Scandinavica*, **80**, 833–6.
Lucas, A. (1991). Programing by nutrition in man. In: *Early Diet, Later Consequences*, ed. D. Conning. *British Nutrition Foundation Nutrition Bulletin*, **16**, Suppl 1, pp. 24–8.
McCance, R.A. & Widdowson, E.M. (1974). The determination of growth and form. *Proceedings of the Royal Society of London*, **185**, 1–17.
Milner, R.D.G. & Hill, D.J. (1987). Interaction between endocrine and paracrine in prenatal growth control. *Journal Paediatrics*, **146**, 113–32.
Smith, C.M. (1994). Food patterns, energy intakes and metabolic demand in children of short stature, children of short stature receiving growth hormone and normal height children. MPhil Thesis, University of Southampton.
Soares, M.J. & Shetty, P.S. (1991). Basal metabolic rates and metabolic economy in chronic undernutrition. *European Journal of Clinical Nutrition*, **45**, 363–73.
Soares, M.J., Piers, L.S., Shetty, P.S., Jackson, A.A. & Waterlow, J.C. (1994). Whole-body protein turnover in chronic undernutrition. *Clinical Science*, **86**, 441–6.
Spurr, G.B. (1988). Effects of chronic energy deficiency on stature, work capacity and productivity. In: *Chronic Energy Deficiency: Consequences and Related Issues*. Ed. B. Schurch & N.S. Scrimshaw. I/D/E/C/G, Nestlé Lausanne, pp. 95–134.
Walker, J.M., Bond, S.A., Wootton, S.A., Betts, P.D. & Jackson, A.A. (1990). Treatment of short normal children with growth hormone – a cautionary tale? *Lancet*, **336**, 1331–4.
Walker, S.P., Powell, C.A. & Grantham-McGregor, S.M. (1990). Dietary intakes and activity levels of stunted and non-stunted children in Jamaica. Part 1, Dietary Intakes. *European Journal of Clinical Nutrition*, **44**, 527–34.
Waterlow, J.C. (1992). *Protein Energy Malnutrition*, pp. 16–20. London: Edward Arnold Publ.
Waters, M.J., Barnard, R.T., Lobie, P.E. *et al.* (1990). Growth hormone receptors –

their structure, location and role. *Acta Paediatrica Scandinavica*, Suppl **366**, 60–72.

Widdowson, E.M. (1962). Nutritional individuality. *Proceedings of the Nutrition Society*, **21**, 121–8.

Widdowson, E.M. & McCance, R.A. (1963). The effect of finite periods of undernutrition at different ages on the composition and subsequent development of the rat. *Proceedings of the Royal Society B*, **158**, 329–47.

Winick, M. & Noble, A. (1966). Cellular response in rats during malnutrition. *Journal of Nutrition*, **89**, 300–6.

8 Early environmental and later nutritional needs

C.J.K. HENRY

Introduction

For over a century, biologists have been interested in the role diet may play in growth and development (Minot, 1891). Whilst recognizing the importance of genetic, environmental and metabolic influences on growth and development (McMeekan, 1940), it was Brody (1945) who developed the theme of 'homeostasis of growth' in his classic '*Bio-energetics and Growth*'. Brody (1926) showed that, when an animal was growth retarded and its restriction removed, the animal grew at a rate greater than was normal in animals of comparable size of age. Some years earlier, Hatai (1906) and Osborne & Mendel (1916) had noted that animals that were re-fed after growth restriction grew faster and 'exceeded the normal growth process', this soon became known as compensatory growth. Compensatory growth after inanition is a common feature in all animals. Normal growth involves the orderly deposition of protein and fat in the body. The growth process can generally be described from two perspectives; first, an increase in body weight with time, commonly described by a growth curve, and secondly, as changes in body composition resulting from differences in the relative growth rates of the component parts of the body.

When an animal is re-fed after inanition, the rate of tissue accretion and organ size development is altered. The question that this chapter aims to address is, does early environmental manipulation in humans lead to long-term alteration in their *nutrient* needs later in life, due to changes in body composition or organ size? Three immediate problems of definition emerge in attempting to address this topic. First, what is meant by the term *early environment*? Which part of the time-frame in the life history of any animals can be considered as 'early environment'? Figure 8.1 shows the life history stages common to most mammalian species. Relative to the extended lifespan of industrialized *Homo sapiens*, it could be argued that 'early environment' could be at any stage in the course of growth and

124

Table 8.1. *Factors influencing growth*

Prenatal	Postnatal
Maternal Age	Nutrition
Maternal nutrition	• Anorexia nervosa
Maternal alcohol consumption	• Atypical eating disorders
Maternal smoking	• Unbalanced diets
Drugs taken during pregnancy	• Psycho-social problems
Order of birth	• Infection related malabsorption
Crowded uterus	• Socio-economic status
Altitude	• Urbanization
Maternal disease (e.g. malaria) and lower birthweight	• Sports activity
Diabetes and higher birthweight	

Conception ➤ Embryo ➤ Fetus ➤ Neonate ➤
Infant ➤ Child ➤ Puberty ➤ Adulthood ➤
Ageing and senescence

Figure 8.1. Life history stages

development. However, Barker and colleagues (1986) have focused specifically on relationships between fetal, neonatal and infant nutrition and chronic diseases of later adulthood. In this chapter, the relationship between environmental manipulation during the neonatal/infant period and adult nutrient needs, is considered.

The second problem is that of defining the nutrient needs of adults. The nutrient needs as widely described are *population* based constructs and should not be used for individuals. The third problem is the definition of 'environment'. Table 8.1 shows the various environmental factors that can influence growth. At the broadest level, there is a prenatal and postnatal environment; the nutrition environment *in utero* is clearly different from the *ex utero* environment and its impact later in life may be markedly different.

Whilst many of these factors are interconnected and have a direct influence on growth, this chapter will concentrate on the *early nutritional* milieu during the neonatal/infant period of the animal, and how this may influence growth, development and later nutrient needs.

The following interconnected issues are examined in this chapter: (i) how does dietary manipulation at various development stages influence growth? (ii) can animals/humans that are growth retarded show compensatory growth on re-feeding? (iii) do the body composition or body compartments in such animals, when re-fed, differ from normal animals with identical body weight? (iv) how might changes in body composition or body compartments influence nutrient needs?

Several difficulties are encountered in any attempt to establish associations between early nutrition in animals/humans and *later growth outcome*. First, the identification of growth outcomes that may be impaired by inadequate food intake *per se* is problematic, since other factors such as infection, intestinal nutrient losses, anorexia, psychosocial pressure can also retard growth. Secondly, humans do not consume nutrients, but eat food within complex ecological and environmental settings. To partition and quantify human diets into precise nutrient intakes is an almost impossible task, especially in the case of retrospective studies (Bingham, 1987). Whilst human diets cannot be easily manipulated within a laboratory setting, animals may be used as models to understand some of the processes of growth retardation during strict nutrient restriction and followed through the phase of compensatory, or catch-up, growth during re-feeding. However, the difficulty arises in the need for animals with similar life-history stages, making primates the most appropriate model. Whilst primate data are limited, other homeotherms whose growth has been manipulated by diet may be used in analysis (Brody, 1945).

Three sources of information are used to explore this theme: (i) experimental studies in animals (ii) studies on malnourished humans responding to nutritional rehabilitation, (iii) analysis of 'natural experiments' (such as human starvation studies) and determining the responses to nutritional rehabilitation.

The period of growth between birth and the onset of puberty is particularly environmentally sensitive (Johnston, Borden & MacVean, 1973). Interestingly, this period of growth in commercially raised animals is of great interest to the meat industry. Hence, a wealth of information exists on how dietary manipulation early in life may influence carcass quality and quantity, and such information forms a useful complement to this investigation. Growth may be considered first as an increase in body size, and secondly, as change in form and body composition. Whilst assessment of body weight is simple, the measurement and evaluation of body compartments (protein, fat, water, carbohydrates and mineral content), and the changes in visceral organ size during growth, is more complex. Such information can only be obtained by subjecting animals and humans to comprehensive anatomical dissection. This has therefore, limited the amount of information available on humans on changes in visceral organ size and body compartment after periods of dietary manipulation.

Table 8.2. *A selection of studies on underfeed-ing in animals*

Investigator	Animal used	Year
Aron	Dogs	1911
Osborne & Mendel	Rats	1915
Jackson	Rats	1920
McMeeken	Pigs	1940
Wallace	Lamb	1948
Wilson	Poultry	1957
Winchester & Ellis	Cattle	1957
McCance &	Rats, Pigs,	1960–
Widdowson	Poultry, Lamb	1968

Experimental studies in animals: refeeding after nutrient restriction

When laboratory and farm animals are deprived of food, their growth slows down. Animals re-fed after transient inanition can achieve normal, presumably genetically maximal body size depending on the timing, duration and stage of the animal's development. Interest in the ability of animals to resume growth after periods of inanition has a long history and may be traced to the earliest study by the German scientist, Aron, in 1911. Aron studied dogs at various stages of inanition in the Philippines. Subsequently, in 1915 the great doyens of nutrition Osborne and Mendal (1916) examined the role of early nutrient restriction on catch-up growth later in life. Table 8.2 shows a selection of studies on animals and the various species used.

Aron's Experiment

Two puppies aged 40 days were brought up under similar condition and at that age their weights were practically the same, about 2 kg. From this time onward one puppy was given a sufficient diet consisting of 400 to 680 Calories per day, while the other received 110 to 175 calories a day. The former increased from 2 to 6.4 kg in this interval of 310 days, but the other retained practically an unchanged weight of 2.4 kg. Although the dog weighing 2.4 kg gained weight on re-feeding, after 310 days on a restricted diet, it did not attain the final weight of the animal fed normally.

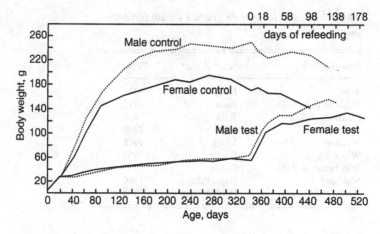

Figure 8.2. Chart showing curves of growth in albino rats amply re-fed after underfeeding from 3 weeks to 1 year of age (Jackson & Stewart '20). The test rats fail to recover fully, remaining permanently stunted.

Osborne and Mendel study

Specific nutrients, such as amino acids, when excluded from the diet can also retard growth. This is shown by examining the classic study of Osborne and Mendel (1916). These investigators used zein protein from corn which was deficient in the essential amino acids lysine and tryptophan. The animals were fed a zein protein for 200 days, when severe growth retardation was observed. However, when the essential amino acids, lysine and tryptophan were added to the diet, the resumption of rapid growth was observed in these animals, even after 200 days of growth restriction.

Jackson Study (1920)

Animals were restricted in food intake from 20 to 342 days. On re-feeding, the animals resumed rapid growth but did not achieve the final weight of the control group (Figure 8.2). Osborne and Mendel (1915) commented on catch-up growth in the following way: 'Growth in the cases referred to, is resumed at a rate normal for the size of the animal at the time. It need not be slow and frequently, it actually exceeds the normal process.'

When animals are restricted in food intake from birth for only 40 days and then re-fed *ad libitum*, large differences in the final weights achieved are also observed (Morgalis, 1923). A similar phenomenon was observed in rats which were diet-restricted from birth until 60 days of age. Two important points emerge from collating information from these studies:

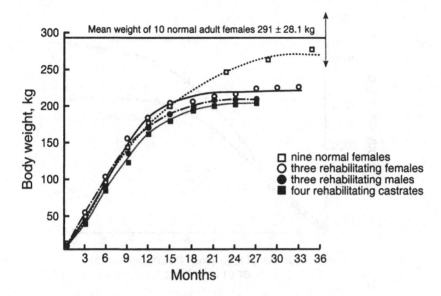

Figure 8.3. Increase in weight of pigs rehabilitated after 1 year of undernutrition and of normal pigs being reared from the same initial weight.

1. Whilst the body appears to exhibit enormous plasticity for recovery after inanition, if the restriction is too severe, subsequent development of body size may be permanently reduced.
2. If food restriction is presented at an early stage of development (even for a short period), recovery on re-alimentation may be incomplete. This concept was eloquently described by McCance and Widdowson in the early 1970s as the 'critical period of growth'.

To illustrate the importance of the timing of inanition, on the subsequent ability of an animal to completely recover its body weight, two experiments conducted by Lister and McCance (1966, 1967) in poultry and pigs can be examined. In the poultry experiment, the animals were kept for 6 months on a restricted intake, followed by food consumption *ad libitum*. This did not lead to complete body weight recovery. Similarly, in the pig experiment, the animals were undernourished for 1 year, then re-fed. These animals also did not achieve full recovery (Figures 8.3 and 8.4).

An interesting feature of these experiments was that when the nutrient requirements for growth in the pig were in short supply, the bones and bone growth took priority in the allocation of nutrients. In other words, bone growth proceeded at the expense of lean or fat tissue accretion.

In summarizing this section, six main factors govern an animal's ability

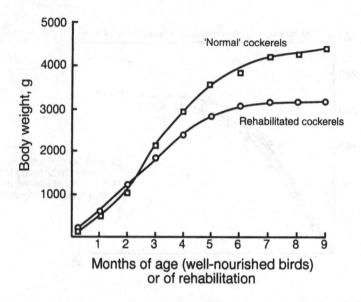

Figure 8.4. Increase in body weight of 'normal' Light Sussex cockerels and of birds of the same breed being rehabilitated after 6 months undernutrition.

to recovery after undernutrition: (i) the nature of undernutrition; (ii) the severity of undernutrition; (iii) the duration of undernutrition; (iv) the stages of development at the time of undernutrition; (v) the relative rate of maturity of the animals; (vi) the pattern of re-feeding.

Weight gain and body composition after re-alimentation

Since weight measurements are the simplest to measure, they have come to play a disproportionately important role in the minds of many investigators. However, weight measurements alone can be misleading. The body is the sum of its parts. Hence, its final size may be limited by the development and growth of one or more of its components. In the 1940s McMeeken, working in Cambridge, made an important observation which has rarely been alluded to. Using pigs and lambs as his experimental animals, he showed that, when animals that were growth restricted early in life, and were rehabilitated to identical body weights as their controls, considerable changes in visceral organ size were seen in animals that were nutritionally deprived early in life. More recently, Koong and Ferrell (1990) have reported a similar study and their experimental scheme is illustrated in Figure 8.5.

Twenty-seven 12 week-old Yorkshire and Landrace cross-bred barrows with an average weight of 27 kg, were used in the study. Three animals from

Table 8.3. *Effect of different nutritional regimes on organ size and heat production in pigs*

	HL	MM	LH	RSD*
Body weight (kg)	40.7	40.9	40.5	1.4
Stomach (g)	263[a]	287[b]	338[c]	17
Small intestine (g)	669[a]	853[b]	1013[c]	98
Large intestine (g)	451[a]	492[b]	588[c]	64
Pancreas (g)	52[a]	63[b]	79[c]	6
Liver (g)	447[a]	537[b]	646[c]	65
Heart (g)	165	157	155	11
Kidneys (g)	112[a]	120[a]	139[b]	11
Spleen (g)	53	51	49	7
Fasting heat production (kcal/day)	1079[a]	1298[b]	1519[c]	110

[abc] Values not showing the same superscript in the same row are significantly different ($P<0.01$).
* Residual Standard Deviation.

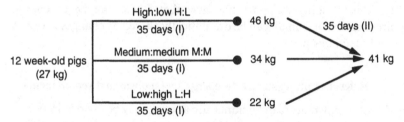

Figure 8.5. Manipulation of organ size (heart, liver, brain, kidneys) in pigs/sheep experimental design of Koong & Ferrell (1990).

each litter were assigned to three treatment groups. The first group, called high–low (H–L), was fed to gain 19 kg during period I, and restricted in food intake in period II. The second group, medium–medium (M–M), was fed to gain 7 kg during period I and II. The third group, low–high (L–H) was diet restricted to lose 5 kg, and re-fed in period II to gain 19 kg. At the end of the experimental period, all the pigs had attained a weight of 41 kg. At the end of the 70-day experimental period, the animals had fasted for 30 hours and fasting heat production was measured for 16 hours by open-circuit indirect calorimetry. The animals were then slaughtered and their organ weights carefully measured. (Table 8.3).

What is clear (Table 8.3) is the considerate reduction in organ weight and size in the animal fed high–low regime compared to the low–high diet regime. Moreover, the fasting metabolic rate in the H–L group was approximately 70% of the value seen in the L–H group, *despite an identical*

Table 8.4. *Organ metabolic rate (OMR) and organ size in adults: relation to BMR*[a]

	Weight (kg)	VO ml/100^{2}g/m)	OMR	OMR/BMR[b] (%)
Brain	1.40	4.2	414	23.2
Liver	1.60	4.1	464	26.1
Heart	0.30	8.2	182	10.2
Kidney	0.30	5.5	116	7.1
Total	3.60		1177	66.7
Skeletal muscle	28.30	0.26	500	28.1
Total	31.90		1677	94.2

[a] Table adapted from Holliday (1971).
[b] BMR for 70 kg man from Talbot (1938).

body weight. This suggests that the visceral organs can be dramatically altered by dietary manipulation. It is well known that the visceral organs make a substantial contribution to basal metabolic rate (BMR) (Holliday, 1971). These findings support the view that BMR and hence energy requirements in an animal, can be dramatically altered by changes in organ size and metabolism.

Relevance of organ size to energy metabolism and requirements

Total energy expenditure in an immature animal, may be divided into four compartments: BMR, diet-induced thermogenesis, growth and activity. BMR usually represents the largest component of energy expenditure (Payne & Waterlow, 1971) and may be defined as the 'energy expended to keep alive'.

In adult humans, it has been shown that four main organs, brain, liver, heart and kidney, which represent approximately 6% of total body weight contribute to 65% of BMR. Muscle mass, on the other hand, which represents up to 30% of *total body weight*, contributes a mere 30% of basal energy expenditure (Table 8.4) (Forbes, 1989).

The pattern of increase in weight of these organs during normal human growth, mirrors impressively the changes in BMR recorded during growth (Holliday *et al.*, 1967) (Figure 8.6). The organ metabolic rates expressed as kcal/kg/day for brain, liver, heart and kidney are as follows: 296, 290, 606 and 388. In contrast, the value for muscle is 17.6. In other words, visceral organ metabolism is 15–25 times the value seen in resting muscle (Holliday, 1971).

These two observations taken together, point to the view that the

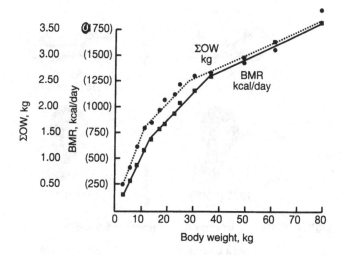

Figure 8.6. The parallel relation between basal metabolic rate (BMR) vs body weight and sum of organ weight (ΣOW) vs body weight from infancy to maturity (males) (Holliday, 1971).

visceral organ makes a major contribution to BMR, and that small weight changes in organ size can therefore make large alterations to BMR.

The most recent FAO/WHO/UNU report (1985) on energy and protein requirements proposed the use of energy expenditure rather than intake as the basis for estimating energy needs in humans. The report also suggested that various components of energy expenditure be expressed as multiples of (BMR). This new approach in estimating energy requirements emphasized the need to assess accurately BMR in populations living under various climatic and environmental conditions. Under- or overestimation of BMR could significantly affect the overall estimation of energy requirements. Whether an individual is asleep, at rest or active, BMR represents a major component of total energy expenditure. Conventionally, BMR is measured using direct or indirect calorimetry. Although it may be accurately measured using these techniques, it is simpler, in practice, to use predictive equations (Harris & Benedict, 1919; Dubois & Dubois, 1916) to estimate BMR and hence energy requirements.

Schofield, Schofield and James (1985) presented predictive equations for both sexes for the following ages: 0–3, 3–10, 10–18, 18–31, 30–60, and >60 years. Their analysis also formed the basis for the equations used by the FAO/WHO/UNU (1985) report on energy and protein requirements. The Schofield analysis and equations were based on a database of 114 published studies of BMR, representing approximately 8000 data points. Their

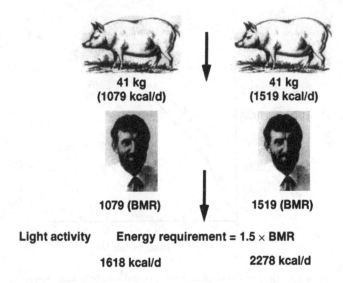

Figure 8.7. Energy requirements in two subjects with identical body weight.

publication is the largest and most comprehensive analysis of BMR to date. Whilst the Schofield equations predicted BMR accurately in many population groups, they were less accurate in predicting BMR in tropical populations (Henry & Rees, 1991; Piers & Shetty, 1993). One possible explanation for this observed lower BMR in tropical people may be the smaller organ weights in these subjects, compared to Caucasian populations.

Relevance of animals studies to humans

It has been shown by Koong and Ferrell (1990) that the organ weight and fasting metabolic rate in animals may be dramatically altered by dietary manipulations. It is tempting to speculate that a similar phenomenon may also occur in humans. Figure 8.7 shows a calculation on a human subject of identical weight, who has a markedly different BMR and hence total energy requirements, merely due to small differences in organ weight. It can be speculated that such differences in organ weight can originate as a consequence of nutritional insults early in life. To put such speculation on firmer footing, it is important to collate information on organ weights in different populations to support this proposition. However, there are many technical and practical limitations associated with obtaining organ weights. Nevertheless, a preliminary analysis of organ weight in different populations is promising.

Table 8.5. *Comparison of male organ weights in different ethnic groups*

Organ	Indian[1] Number of subjects	Average weight (g)	Burmese[2] Number of subjects	Average weight (g)	African[3] Number of subjects	Average weight (g)	Caucasian[4] Number of subjects	Average weight (g)
Liver	207	1273	100	1301	100	1471	79	1624
Heart	112	198	100	283	100	254	79	338
Kidney	120	226	100	283	100	254	79	301
Brain	17	1273	100	1330	100	1216	79	1395
Total		2970		3197		3141		3658

[1] Buchnan and Daly (1902).
[2] Castor (1912).
[3] Castor (1912).
[4] Greenwood and Brown (1913).

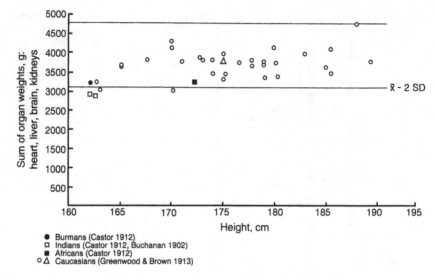

Figure 8.8. Sum of organ weights in male subjects (Greenwood & Brown, 1913).

Organ size from different ethnic groups were obtained from Castor (1912) and Greenwood and Brown (1913); these substantial data-sets are shown in Table 8.5. It is clear from the Table that the *average* visceral organ weight of Indian subjects was small compared to other ethnic groups. Figure 8.8 shows a graphical representation of the results, with the mean – 2SD value for the organ weight in Caucasian subjects. It is interesting to note that the Indian value falls just below the mean – 2SD line. Whilst these results must be treated as being preliminary, they provide a rational basis for the widely reported observation that the BMR in Indian subjects is lower than Caucasian subjects (Mukerjee & Gupta, 1931) largely due to differences in organ size.

Conclusions

A survey of animal studies, coupled with some human observations, point to the view that early nutrition may influence organ size and hence energy requirements. Thus, early nutritional environment may play a role in later nutrient needs from the evidence preserved here, at least in determining energy requirements.

It is tempting to propose that the reduced BMR in certain populations is due to smaller organ size, similar to that seen in lambs and pigs. An unknown nutrient may be responsible for the reduction in organ size. Whilst little is known about the nutrients required for linear growth

(Waterlow, 1988; Golden & Golden, 1981), even less is known about the nutrient needs for organ growth/development. In humans, the period between birth and the first few years of life may be most sensitive to nutritional deprivation which might lead to an alteration in visceral organ size development, and the relative contribution of visceral organs to overall body composition.

Whilst this chapter may have posed more questions than it answers, it highlights the need to examine the relationship between organ size, energy metabolism and energy needs. The relationship between body size and energy need is not a simple one, and the role that organ size may play in determining energy needs cannot be ignored.

References

Aron, H. (1911). Nutrition and growth. *Philippine Journal of Sciences, Section B (Medical Science)*, **6**, 1–52.

Barker, D.J.P., Osmond, C. (1980). Infant mortality, childhood nutrition and isochemic heart disease in England and Wales. *Lancet*, **i**, 1077–81.

Bingham, S.A. (1987). The dietary assessment of individuals. *Nutrition Abstracts and Reviews*, **57**, 705–42.

Brody, S. (1926). Growth and development, with special reference to domestic animals. *Research Bulletin Missouri Agricultural Experimental Station*, 103.

Brody, S. (1945). *Bioenergetics and Growth*. New York: Reinhold.

Castor, R.H. (1912). Weight of organs. *Journal of Tropical Medicine and Hygiene*, **15**, 17–24.

Dubois, D. & Dubois, E.F. (1916). A formulae to estimate the approximate surface area. *Archives of Internal Medicine*, **17**, 863–71.

FAO/WHO/UNU (1985). Energy and protein requirement. *Technical Report Series*, **724**. Geneva: WHO.

Forbes, G.B. (1989). Maintenance of energy needs for women as a function of body size and composition. *American Journal of Clinical Nutrition*, **50**, 404–5.

Golden, M.H.N. & Golden, B.E. (1981). Effect of zinc supplementation on the dietary intake, rate of weight gain and energy cost of tissue deposition in children recovering from severe malnutrition. *American Journal of Clinical Nutrition*, **34**, 900–8.

Greenwood, M. & Brown, J.W. (1913). A second study of the weight, variability and correlation of the human viscera. *Biometrica*, **9**, 473–85.

Hamilton, T.S. (1939). The growth, activity and composition of rats fed diets balanced and unbalanced with respect to protein. *Journal of Nutrition*, **17**, 565–82.

Harris, J. & Benedict, F.G. (1919). *A Biometric Study of Basal Metabolism in Man*. Carnegie: First Washington, 79.

Hatai, S. (1906). Effect of partial starvation followed by return to normal diet. *American Journal of Physiology*, **18**, 309–20.

Henry, C.J.K. & Rees, D.G. (1991). New Predictive equation for the estimation of basal metabolic rate in tropical peoples. *European Journal of Clinical Nutrition*, **45**, 177–85.

138   C.J.K. Henry

Holliday, M.A., Potter, D., Jarrah, A. & Bearg, S. (1967). Relation of metabolic rate to body weight and organ size. *Paediatric Research*, 1, 185–92.

Holliday, M.A. (1971). Metabolic rate and organ size during growth from infancy up to maturity and during late gestation and early infancy. *Paediatrics*, 47, 169–79.

Horst, K., Mendel, L.B. & Benedict, F.G. (1930). The metabolism of the albino rat during prolonged fasting at two different environmental temperatures. *Journal of Nutrition*, 3, 177–200.

Jackson, C.M. & Stewart, C.A. (1920). Effect of inanition in the young upon the ultimate size of body. *Anatomical Records*, 3, 97–128.

Johnston, F.E. Borden, M. & MacVean, R.B. (1973). Height, weight and their growth velocities in Guatemalan private schoolchildren of high socio-economic class. *Human Biology*, 45, 627–41.

Koong, L.J. & Ferrell, C.L. (1990). Effect of short term nutritional manipulation on organ size and fasting heat production. *European Journal of Clinical Nutrition*, 44, suppl. 73–7.

Lister, D., Cowen, T. & McCance, R.A. (1966). Severe under nutrition in growing and adult animals. *British Journal of Nutrition*, 20, 633–9.

Lister, D. & McCance, R.A. (1967). Severe under nutrition in growing and adult animals. *British Journal of Nutrition*, 21, 787–97.

McMeekan, C.P. (1940). Growth and development of the pig, with special reference to carcass quality. *Journal of African Science*, 31, 1–49.

Minot, C.S. (1891). Senescence and rejuvenation. *Journal of Physiology*, 1, 3–9.

Morgulis, S. (1923). *Fasting and Undernutrition*. New York: Dutton.

Mukerjee, H.N. & Gupta, P.C. (1931). The basal metabolism of Indians (Bengalis). *Indian Journal of Medical Research*, 11, 807–11.

Osborne, T.B. & Mendel, L.B. (1916). Acceleration of growth after retardation. *American Journal of Physiology*, 40, 16–20.

Payne, P.R. & Waterlow, J.C. (1971). Relative requirements for maintenance, growth and physical activity. *Lancet*, ii, 210–11.

Piers, L.S. & Shetty, P.S. (1993). Basal metabolic rate of Indian women. *European Journal of Clinical Nutrition*, 47, 586–91.

Schofield, W.N., Schofield, E. & James, W.P.T. (1985). Predicting basal metabolic rate. *Human Nutrition and Clinical Nutrition*, 39c, Suppl., 5–41.

Talbot, F. (1983). Basal metabolism standards for children. *American Journal Diseases of Childhood*, 55, 455–9.

Waterlow, J.C. (1988). *Linear Growth Retardation in Less Developed Countries*. Nestlé Nutrition seminars 14. NY: Raven Press.

Wilson, P.N. (1957). Effect of contrasted planes of nutrition on the carcass composition of the East African Dwarf goat. *Nature, London*, 180, 145–6.

9 Ontogeny of human taste and smell preferences and their implications for food selection

DAVID J. MELA AND SUSAN L. CATT

Introduction

It is commonly recognized that humans raised in a given cultural environment prefer customary foods and flavour combinations. Along with geography and economics, this is perhaps the predominant determinant of the range of foods accepted and eaten amongst a given population. Within a cultural or family environment, though, different individuals also display varied likes and dislikes for particular foods. Possible underlying explanations for the emergence of similarities and differences amongst individuals can be drawn from both the social and biological sciences.

For individuals within a culture it is often assumed, without clear evidence, that lasting taste and smell preferences may develop early in life, and have a major influence on later food selection and intake. Following from this, it is plausible that this may influence body size, composition, and metabolic characteristics of populations and their subsets (Wootton & Jackson, this volume). However, it is extraordinarily difficult to define where taste and smell preferences play a primary role in food acceptance, and where their role is secondary to or determined by individual and cultural attitudes, beliefs, and norms. A categorization described by Rozin (1989; Rozin & Fallon, 1981) may be useful here for classification of the various influences on food likes and dislikes, and clarifying the present discussion. Using his headings, the 'sensory-affective' dimension relates to acceptance or rejection based on hedonic value of the sensory quality, including taste and smell. That is, actual liking for the specific flavour, smell, and texture of the food, apart from consideration of its origin, appropriateness, familiarity, potential after-effects, or other characteristics. 'Anticipated consequences' refers to the perceived after-effects of consumption of a food, ranging from short- or long-term impacts on health, to social standing and prestige. Foods may be accepted or rejected because they are viewed as potentially harmful or beneficial in some way, reasons which may be largely unrelated to sensory qualities. Lastly, 'ideational' factors reflect what the food is, its origin, or its cultural or symbolic meaning, not specifically linked to the potential consequences of eating it. Rozin & Fallon (1981) give a thorough description of these categories, with examples and potential implications for food choice.

This review will attempt to focus on the 'objective' sensory-affective dimension of food acceptance. It is critical, however, to reiterate the point that food acceptance frequently has little to do with sensory-affective judgements, even if that is not immediately apparent to the consumer. Foods are often rejected or accepted for reasons largely unrelated to their sensory characteristics. Ideational factors, for example, have an important influence in determining what constitutes the food domain within a culture, and (with further subdivisions of 'inappropri-ate' and 'disgusting') have a major role in the rejection of certain foods and food combinations. Items or contexts which might be viewed as inappropriate or disgusting are typically regarded *a priori* as unpalatable (Rozin & Fallon, 1981). The reasons that the British do not eat pickled beetroot for breakfast, peanut butter on peas, or slugs at any time arguably have little to do with sensory affect specifically; indeed, few of these will ever have been tried. Learning about appropriateness and context takes place early in life (see Birch, 1991), and these psychosocial influences can undoubtedly be a powerful and lasting force in the early learning environment and in subsequent food beliefs and acceptance. Taste and smell operate interactively amongst the myriad influences on the selection and consumption of food by children and adults (Ray & Klesges, 1993; Shepherd, 1989).

This chapter will briefly review the development of taste and smell in humans, focusing on the 'innate' and learned influences on sensory likes and dislikes, and consider the evidence for their possible relationships to later food selection. The focus will be on human research, supplemented by reference to the much larger animal literature where it provides useful supporting evidence or additional issues.

Characteristics of chemosensory systems

Perceptions of the sensory characteristics of foods are largely derived from combinations of sensations mediated via the gustatory, olfactory, and trigeminal systems. These systems are briefly characterized in Table 9.1. It should be made abundantly clear that these peripheral sensory systems are largely distinct in anatomy and physiology, as well as in the sensations they

Table 9.1. *Sensory systems involved in flavour perception*

Sensory system	Sensations mediated
Gustatory	Taste: sweet, sour, salty, bitter, others? ('umami'?)
Olfactory	Odour: unlimited number?
Trigeminal	Tactile, thermal, pain/irritation

mediate. Nevertheless, confusions in the meaning and function of 'taste' and 'smell' are frequent even in the scientific literature.

Although the word 'taste' is commonly equated with 'flavour' in colloquial speech, the former technically describes only the sensations derived from the limited repertoire of the gustatory system, and that is how the term will be used here. The predominant view that taste is limited to mixtures of four basic qualities, sweet, sour, salty, bitter, is supported by a number of lines of evidence (see McBurney & Gent, 1979). In addition, there is increasing agreement that the savoury quality associated with MSG and ribonucleotides merits general acceptance as a basic taste ('umami'), as it is in Japanese science and culture. Nevertheless, there is still some debate: other schemes have been proposed, and it is suggested that the 'basic' tastes may merely be easily recognized points in a continuum, or the result of limitations in semantics or cultural exposure. However, biochemical evidence increasingly supports the concept of four or five discrete categories of taste quality.

'Flavour' is generally used to describe the overall chemically elicited sensation of the food, including inputs from all the systems noted in Table 9.1. More specifically, it is used to refer to volatile flavours, i.e. odours or aromas. A major functional distinction between taste and smell is that, while many foods share common taste characteristics (e.g. sweetness, sourness, etc.), it is in their volatile flavours that individual foods and other objects may be recognized and identified as unique (many fruits taste sweet, but only one has the flavour of a strawberry). Thus, there are thousands of individually discernible odours and, despite many attempts, there is not an agreed classification scheme for categorization of 'primary' or 'basic' odour qualities. As will be described, this diversity and specificity has critical implications for development of learned associations.

Unlearned ('innate') sensory responses

It should be noted that use of the word 'innate' to characterize sensory responses present at birth and/or universal in humans may be confusing. Such 'innate' responses are not necessarily genetically endowed, and may in many cases be best explained as a function of *in utero* experience, although some aspects of that may be common to all humans. The distinction here will be made between those responses which are likely to be unlearned, against those acquired through experience.

Taste preferences

There have been a large number of studies examining various aspects of responsiveness to basic chemosensory stimuli (simple tastes and odours) by newborn infants. These indicate that humans are born with positive hedonic responses to at least one sensory quality, sweetness, and probably a dislike for bitter and sour tastes. Ability to sense (and like) salty stimuli probably develops postnatally in humans, and there may also be some postnatal maturation of bitter taste perception. These 'innate' taste preferences are characterized in Table 9.2, and the lines of support for them are thoroughly discussed in Beauchamp, Cowart & Schmidt (1991). It is logical to presume that the normal neonatal hedonic responses to these basic taste qualities are truly unlearned: the types of stimuli which give rise to these taste sensations are shared by many common biological compounds, there are relatively specific receptor elements and/or transduction mechanisms for each of these taste qualities, and related animal species show hedonic responses broadly similar to those of humans. That does not, however, exclude the possibility that these innate responses might be modified by later experience and learning.

Table 9.2. *Unlearned hedonic responses to taste qualities*

Quality	Accept or reject	Unlearned?	Development
Sweet	Accept	Yes	*In utero*
Salt	Accept	Apparently	4–6 months postnatal
Sour	Reject	Probably	Present at birth (?)
Bitter	Reject	Apparently	*In utero* (?) + postnatal maturation

Teological explanations have often been put forward to account for these apparently universal human chemical sensitivities and preferences. Usually, the liking for sweetness is attributed to the fact that naturally occurring sweeteners are associated with safe sources of energy and selected nutrients (see Ramirez, 1990 for a critical analysis of this point). Salt taste is suggested to promote ingestion of sufficient sodium and, perhaps, other minerals. Many natural toxicants are bitter-tasting, and it is widely accepted that bitter taste perception confers the capacity to recognize and avoid such substances. Lastly, the sensitivity to pH represented by sour taste is an extremely primitive and ubiquitous characteristic of living cells.

Odour and food preferences

Far less is established regarding possible unlearned preferences for specific odours and foods. Compared to studies of taste, relatively little work has been carried out on volatile odours or more complex systems or food models, in part due to problems of stimulus control and constraints on the types of materials to which infants may be exposed. Furthermore, as noted, there are no discrete, agreed categories of odours from which representative compounds might be easily selected for testing. Evidence for universal or unlearned odour preferences is lacking. Although liking of food-related and other aromas appear to show, with some exceptions, broad general similarities across different human populations (Schleidt, Neumann & Morishitra, 1988; Pangborn, Guinard & Davis, 1988), this may better reflect shared experiences with common environmental odours than a specific genetic predisposition. The work of Davis (see Story & Brown, 1987) may also be relevant to consideration of unlearned food preferences, as it is frequently cited as indicating that children have an instinctive capacity to choose a nutritionally adequate diet. However, that simplistic interpretation of Davis' work has been challenged (Story & Brown, 1987), and Galef (1991) has also questioned the general notion that humans and other omnivores have an inborn drive to self-select appropriate diets. While young infants rapidly learn to recognize and respond to selected environmental odours, recent work has suggested a possible inborn attraction of neonates to the odour of lactating breasts (Porter *et al.*, 1991; discussed below).

Genetic factors are known to markedly influence the ability to sense certain specific taste and odour compounds (see Beauchamp & Wysocki, 1990; Drewnowski, 1990; Pangborn, 1981), but much less is known about the heritability of taste and smell preferences. Rozin (1991*a*) has characterized the incongruous literature on parent–child resemblances in food preferences as the 'family paradox'. As noted, identifiable cultural food preferences, largely conveyed by parents, clearly exist; the paradox appears when one moves into a finer analysis, where variations in food preferences within a culture are characterized by surprisingly low parent–child correlations (see Rozin, 1991*a*; Borah-Giddens & Falciglia, 1993). The strongest relationships have been shown amongst siblings, where Pliner & Pelchat (1986) report a correlation of $r = 0.50$. Studies of twins suggest that genetic factors play an extremely small part in family resemblances in food preferences (Green, Desor & Maller, 1975; Rozin & Millman, 1987; Krondl *et al.*, 1983), although a more recent study of adult twins (De Castro, 1993) points to significant genetic influences on the selection and intake of certain beverages, including alcohol and coffee. However, pending further study, the present consensus is that genetic factors and family influence account for only a small part of variability in food preferences.

Learned sensory responses

The preceding discussions should serve to indicate that humans have a very limited battery of inborn, universal sensory preferences. A large component of likes and dislikes for the sensory characteristics of specific foods must be acquired through shared or differential experiences. The specific bases for these individual preferences are extremely difficult to unravel, given the attitudinal, social, and psychological contexts of food consumption, and the fact that humans and other animals, including invertebrates, have the capacity to link the consequences of ingestion, metabolic, physiological, psychological, social, to specific foods or to their associated sensory characteristics. Several authors (Rozin, 1989; Rozin & Schulkin, 1990; Birch 1989, 1991; Sclafani, 1990) have written clear, comprehensive reviews of these processes in the acquisition of food preferences, and the interested reader would be well served to refer to those authors.

'Mere exposure' and taste and smell preferences

The concept of becoming 'accustomed to' certain foods and sensory qualities raises the notion that 'mere exposure' to a given sensory quality can result in heightened acceptance of that quality, at least in a given context. However, any discussion of this must acknowledge the theoretical limitations of this process. First, it may be attitudes toward the food, not actual liking for the sensory characteristics of the foods which has changed. Positive attitudes toward a particular food are associated with more positive hedonic ratings in sensory tests (Aaron, Mela & Evans, 1994). Secondly, exposure is rarely 'mere': it is usually conditionally linked to some (though perhaps not obvious) outcome, if not to an identifiable physiological effect.

Taste

There have been numerous animal studies examining how variations in simple exposure at different points in development might influence subsequent preferences for materials representing the basic tastes. These studies have, however, led to contrasting results. At one extreme, Hill and colleagues (Hill, Mistretta & Bradley, 1986; Hill & Przekop, 1988; Vogt & Hill, 1993) have shown that there is a critical period in early embryonic development during which maternal sodium deprivation may have long-term neural and/or behavioural effects on salt taste. It is important to note the point made by the authors, that this is unlikely to reflect changes in fetal sensory exposure to sodium *per se* during gestation, but more likely results from alterations in maternal hormones or metabolites. Although sodium deficiency of this type is extraordinarily unlikely in humans, this is an apparent example of a long-term effect of fetal environment on sensory physiology.

There is a moderate level of evidence that human hedonic responses to basic tastes may be modified by extended sensory experience from dietary exposure. The best controlled examples in humans come from research on salt taste (e.g. Bertino, Beauchamp & Engelman, 1982, 1986; Blais *et al.*, 1986). These studies indicate that the most preferred level of salt in foods can be reduced by about 8 to 12 weeks of consumption of a low sodium diet, and may increase when foods are chronically supplemented with salt. This appears to be a phenomenon related to sensory exposure, and not a physiological effect of the modified sodium intake *per se* (Beauchamp & Engelman, 1991). Sullivan & Birch (1990) showed that young children readily learn about appropriate taste context for a given food through repeated exposure. They gave 4 and 5 year-old children either the sweetened, salted, or plain version of a novel food (tofu) 15 times over 9 weeks. The children not only increased their liking for the specific version to which they had been exposed, but also exhibited a relative decrease in liking for the alternative versions.

There is also some documentation of shifts in taste preferences as a result of actual dietary habits. Beauchamp & Moran (1982) and Harris & Booth (1985) have reported heightened ingestive responses to sucrose- and salt-containing stimuli, respectively, amongst 6 months-old infants in relation to history of dietary experiences with sweetened water and salted foods. Evidence for racial and ethnic differences in hedonic responses to simple taste solutions is found in some, but not all studies (Moskowitz *et al.*, 1975; Greene *et al.*, 1975; Beauchamp & Moran, 1982; Bertino, Beauchamp & Jen, 1983; Bertino & Chan, 1986; Desor & Beauchamp, 1987; Druz & Baldwin, 1982; Prescott *et al.*, 1992). An extreme but frequently cited example is the description by Moskowitz *et al.* (1975) of a group of Indian labourers who, by comparison to western and other Indian subjects, exhibited marked positive hedonic responses to sour and bitter solutions. Such cultural differences are typically viewed as a result of customary dietary context, in this latter case reflecting habitual consumption of the very sour tamarind fruit.

Foods and flavours

For reasons noted previously, differences in sensory hedonic responses to food-related aromas and whole foods are far more complex, would be much more likely to shift with experience, and would be anticipated to display much greater individual and cultural diversity than simple tastes. Birch and colleagues have carried out an extensive range of experiments demonstrating that food preferences of infants and young children can be readily manipulated through repeated exposure (e.g. Birch & Marlin, 1982; for review see Birch, 1989, 1991). These studies and others in adults (Pliner, 1982) have demonstrated relatively clear and robust relationships between exposure to novel foods and their acceptance. In addition, the discussion below will describe how exposure to flavours *in utero* may enhance their acceptance.

As noted initially, the mechanisms associated with the process of 'mere exposure' are not clear. Children (and adults) presumably overcome an inherent neophobia (see Birch & Marlin, 1982), form associations between ingestion of a food and its positive (or absence of negative) consequences, and may shift their beliefs and attitudes regarding the appropriateness and acceptability of the foods or their contexts. Rozin (1991b; Rozin, Gruss & Berk, 1979) has written extensively on the liking for the burn of hot peppers, which illustrates the difficulty of assigning causality even to rather extreme examples of acquired food preferences.

Conditioned sensory-affective responses

Humans are endowed with the ability to recognize and learn from the contexts and consequences associated with ingestion of specific foods or sensory qualities (see Booth, 1985; Rozin, 1989; Rozin & Schulkin, 1990), and many examples of human preferences and aversions for specific foods can be best explained by such processes. These types of learned hedonic responses are often highly robust, can supersede innate preferences, and may be extremely stable. Two features of this process are notable: First, it produces a true shift in perceived sensory hedonic quality of a stimulus. That is, the sensory quality itself (e.g. flavour) becomes more or less liked, apart from (or in spite of) pre-existing attitudes and beliefs regarding the food itself. Secondly, this is not a conscious process. Although one might retrospectively assign an explanation, the shift in liking is not subject to voluntary control.

Conditioned aversions

Intense dislikes or aversions to specific sensory qualities or foods can be rapidly acquired via association with negative outcomes, especially nausea. (Alcoholic beverages present an apparent contradiction to this, which can be explained in a number of ways [see Forbes & Rogers, 1994].) In the classic example, the aroma, flavour and ingestion of a certain food item becomes repulsive, even though the item may have previously been well liked. The formation of aversions is most commonly linked to nausea, and dislike of a food can remain long after the initial incident is forgotten, often despite the knowledge that the item is known to be safe to eat or did not cause the illness. These types of aversions are prevalent in humans, and may be directed at a wide variety of common foods (Garr & Stunkard, 1974; Mattes, 1991). In situations where adults can recall the incident associated with the initial formation of an aversion, it often has a long history, perhaps several decades, and may be identified as having been acquired during childhood. Thus, food aversion learning is a powerful phenomenon and can account for an intense dislike of selected sensory attributes, usually flavours of particular foods.

Conditioned preferences

In a similar manner, albeit less readily, preferences can be fostered by pairing specific qualities with positive physiological or psychological outcomes. It is more difficult to produce convincing examples of conditioned preferences in humans, since the process is much more gradual and less accessible to conscious association. It seems reasonable to ascribe preferences for (innately disliked) items such as coffee and alcoholic beverages to their psychobiological effects (e.g. Rogers & Richardson, 1993). The apparent human preference for the sensory properties of fats in foods also cannot be related reliably to any known innate mechanisms; the available evidence suggests that this is secondary to their association with the postingestive effects of fats (see Johnson, McPhee & Birch, 1991; Kern et al., 1993; reviews by Mela, 1990, 1992).

There is abundant evidence for the predominance of these and other learning processes in the formation of sensory preferences. Numerous experiments in animals have shown how the physiological consequences of foods, including subtle nutritional and metabolic effects, can markedly alter sensory preferences, including reversal of innate likes and dislikes.

Flavour preferences in humans have been shown to be influenced by the energy content (as fat or carbohydrate) of conditioning media (Booth, Mather & Fuller, 1982; Birch et al., 1990; Johnson et al., 1991; Kern et al., 1993). The recent studies of Birch and colleagues (Birch et al., 1990; Johnson et al., 1991; Kern et al., 1993) with children are particularly clear and convincing demonstrations of these processes. In these studies, manipulation of sugar content in beverages or fat in yoghurts effected shifts in flavour preferences; specifically, there was a relatively increased preference for those flavours associated with fat or sugar. The careful counterbalancing, sample preparation, and sensory pretesting procedures make a strong case for attributing the outcome to the metabolic or other reinforcing effects of fat or sugar, and not simply a result of sensory differences in the samples. Such studies convincingly demonstrate that sensory acceptance can be readily shifted by association with differential nutritional and physiological properties of foods. These phenomena may have implications for long-term acceptance of, for example, reduced-energy products, which disengage the traditional linkage of sensory and psychobiological or metabolic associations in foods.

Less is known about the possible influence of the psychobiological effects of foods, (e.g. on mood and performance), and the role of situational and environmental variables as mediators of conditioned sensory preferences. Recent advances in testing of mood and cognitive performance may allow for identification of such effects. Many 'idiosyncratic' differences and shifts in likes and dislikes could presumably be accounted for by associations with these psychobiological properties of foods or their interaction with the eating environment (Rogers & Richardson, 1993; Rogers & Lloyd, 1994).

In utero *and neonatal learning*

There is large body of evidence to suggest that sensory preferences may be shifted through exposure to flavours *in utero* and in mother's milk. Schaal and Orgeur (1992) have presented a concise and convincing review of evidence for olfactory function and learning *in utero*, and cite an extensive body of evidence in support of the following requirements for transnatal chemosensory learning:

- Maturation of chemoreceptor systems *in utero*
- Stimuli efficiently transferred into fetal environment
- Chemosensory memory carries into postnatal period
- Salient stimuli present in both pre- and postnatal periods
- Neonatal detection of stimuli experienced *in utero*

They cite a number of experimental animal studies in which mere exposure effects and conditioned preferences and aversions have been manifested postnatally following exposure of animals to flavours *in utero*, and their analysis strongly suggests that these phenomena should also occur in humans. These may also serve an important biological function, as odours associated with amniotic fluid appear to have a critical role in mother–infant recognition and bonding and for nipple seeking and attachment behaviours (Poindron et al., 1993). It has long been known that food-associated flavours are transmitted into the milk of lactating farm and laboratory animal species. Influences of maternal diet flavour during gestation and lactation on weaning food preferences are well established

(Galef & Henderson, 1972; Galef & Sherry, 1973), and applications of this have been described in the animal production literature (e.g. Rossi, 1980).

Human infants clearly have the capacity for olfactory learning at birth, and a preferential recognition or response of human infants toward natural and artificial odours of their mothers or carers has been demonstrated many times (see Porter *et al.*, 1991 for discussion). Sullivan *et al.* (1991) have shown that 1 day-old human infants can be conditioned to respond selectively to specific odours. This memory may also be retained over a significant period of time. Davies and Porter (1991) exposed newborns to a specific odour for 22 hours within 2 days of birth, and observed preferential orientation towards that odour 2 weeks later.

This type of learning about relevant environmental odours would be expected to readily extend to food-associated flavours. Porter *et al.* (1991) examined the time that 2 week-old formula-fed infants oriented towards test odours from different sources, held to either side of their head. As might be anticipated, infants spent more time orientated towards their mother's breast pads than towards clean control pads. However, males also spent markedly more time orientated toward pads containing their formula odour than towards their mother's breast pads (no difference for females). This is presumably a conditioned preference for the formula odour. Perhaps most surprisingly, formula-fed infants of both sexes spent more time oriented toward the breast odour of an unrelated lactating mother than toward the odour of their own formula. This latter result suggests an inborn attraction to an odour of human milk or lactating breasts; though it does not discount the possibility that the response reflects an outcome of learning which occurred *in utero*.

There are frequent anecdotal reports of flavours and other substances influencing the characteristics of human breast milk and infant behaviour (Font, 1990; Wallace, Inbar & Ernsthausen, 1992), although more objective confirmation of this has only recently appeared. Mennella and Beauchamp (1991*a,b*) have shown that garlic and alcohol consumption influence both the sensory qualities of human milk and the nursling's behaviour. When milk smelled of garlic (as confirmed by an adult sensory panel), infants attached to the breast for longer periods of time and there was some tendency toward increased milk ingestion. In contrast, infants consumed a significantly lower volume of milk following maternal alcohol consumption. Subsequent work has verified these findings, while showing that these effects of garlic attenuate with repeated exposure (Mennella & Beauchamp, 1993*a,b*).

The work of Menella & Beauchamp suggests potentially important implications of human infant exposure to flavours in breast milk. Sullivan and Birch (1994) have reported differences between breast and formula-fed infants in their acceptance of a new solid weaning food (peas and green beans). Although both breast- and bottle-fed infants had comparable initial ingestive responses to the vegetables, breast-fed infants were consuming significantly more vegetables following repeated exposure. It therefore seems likely that, as shown in other animals, acceptance and consumption of new foods by human infants may be facilitated by prior exposure to sensory cues in mother's milk.

Conclusions and long-term implications

It is commonly assumed that taste, smell, and food preferences acquired in infancy are in some way related to taste, smell, and food preferences exhibited in childhood, and these in turn to adult eating behaviour. Unfortunately, the lack of longitudinal studies of chemosensory function and hedonics makes it virtually impossible to verify these assumptions with any degree of confidence. Indeed, much of the available evidence discussed here suggests exactly the opposite; i.e. humans have few 'hard-wired' food preferences. Taste and smell preferences are highly dynamic, and responsive to differences and shifts in their associations with particular social–environmental contexts and psychobiological effects. With the exception of self-histories of food aversions, there is no evidence that specific taste and smell preferences formed early in life actually track into adulthood. On the other hand, many seemingly idiosyncratic food preferences of individual adults could have their roots in early childhood, suckling or even fetal experiences. Unfortunately, current knowledge is too limited to establish this. As Rozin (1991a) concluded, 'If one is looking for variance to explain, the domain of preferences still offers a fertile field'.

References

Aaron, J.I., Mela, D.J. & Evans, R.E. (1994). The influences of attitudes, beliefs and label information on perceptions of reduced-fat spread. *Appetite*, 22, 25–37.

Beauchamp, G.K., Cowart, B.J. & Schmidt, H.J. (1991). Development of chemosensory sensitivity and preference. In *Smell and Taste in Health and Disease*, ed. T.V. Getchell, R.L. Doty, L.M. Bartoshuk and J.B. Snow Jr., pp. 405–16. New York: Raven Press.

Beauchamp, G.K. & Engelman, K. (1991). High salt intake. Sensory and behavioral factors. *Hypertension*, 17, 176–81.

Beauchamp, G.K. & Moran, M. (1982). Dietary experience and sweet taste experience in human infants. *Appetite*, 3, 139–52.

Beauchamp, G.K. & Wysocki, C.J. (1990). Perception of the odor of androstenone:

Influence of genes, development, and exposure. In: *Taste, Experience, and Feeding*, ed. E.D. Capaldi and T.L. Powley, pp. 105–15. Washington, DC: American Psychological Association.

Bertino, M., Beauchamp, G.K. & Engelman, K. (1982). Long-term reduction in dietary sodium alters the taste of salt. *American Journal of Clinical Nutrition*, **36**, 1134–44.

Bertino, M., Beauchamp, G.K. & Engelman, K. (1986). Increasing dietary salt alters salt taste preference. *Physiology and Behavior*, **38**, 203–13.

Bertino, M., Beauchamp, G.K. & Jen, K.C. (1983). Rated taste perception in two cultural groups. *Chemical Senses*, **8**, 3–15.

Bertino, M. & Chan, M.M. (1986). Taste perception and diet in individuals with Chinese and European ethnic backgrounds. *Chemical Senses*, **11**, 229–41.

Birch, L.L. (1989). Developmental aspects of eating. In: *Handbook of the psychophysiology of human eating*, ed. R. Shepherd, pp. 179–203. Chichester UK: Wiley.

Birch, L.L. (1991). Children's experience with food and eating modifies food acceptance patterns. In *Chemical Senses. Volume 4: Appetite and Nutrition*, ed. M.I. Friedman, M.G. Tordoff & M.R. Kare, pp. 283–302. New York: Marcel Dekker, Inc.

Birch, L.L. & Marlin, D.W. (1982). I don't like it, I never tried it: effects of exposure on two-year-old children's food preferences. *Appetite*, **3**, 353–60.

Birch, L.L., McPhee, L., Steinberg, L. & Sullivan, S. (1990). Conditioned flavor preferences in young children. *Physiology & Behavior*, **47**, 501–5.

Blais, C.A., Pangborn, R.M., Borhani, N.O., Ferrell, M.F., Prineas, R.J. & Laing, B. (1986). Effect of sodium restriction on taste responses to sodium chloride: a longitudinal study. *American Journal of Clinical Nutrition*, **44**, 232–43.

Booth, D.A., Mather, P. & Fuller, J. (1982). Starch content of ordinary foods associatively conditions human appetite and satiation, indexed by intake and eating pleasantness of starch-paired flavors. *Appetite*, **3**, 163–84.

Booth, D.A. (1985). Food-conditioned eating preferences and aversions with interoceptive elements: conditioned appetites and satieties. *Annals of the New York Academy of Sciences*, **443**, 22–41.

Borah-Giddens, J. & Falciglia, G.A. (1993). A meta-analysis of the relationship in food preferences between parents and children. *Journal of Nutrition Education*, **25**, 102–7.

Davis, L.B. & Porter, R.H. (1991). Persistent effects of early odor exposure on human neonates. *Chemical Senses*, **16**, 169–74.

De Castro, J.M. (1993). A twin study of genetic and environmental influences on the intake of fluids and beverages. *Physiology and Behaviour*, **54**, 677–87.

Desor, J.A. & Beauchamp, G.K. (1987). Longitudinal changes in sweet preference in humans. *Physiology and Behavior*, **39**, 639–41.

Drewnowski, A. (1990). Genetics of taste and smell. *World Review of Nutrition and Dietetics*, **63**, 194–208.

Druz, I.L. & Baldwin, R.E. (1982). Taste thresholds and hedonic responses of panels representing three nationalities. *Journal of Food Science*, **47**, 561–9.

Font, L. (1990). 'Incidental' maternal dietary intake and infant refusal to nurse. *Journal of Human Lactation*, **6**, 9.

Forbes, J.M. & Rogers, P.J. (1994). Food selection. *Nutrition Abstracts and Reviews*, **64**, 1065–78.

Galef, B.G. (1991). A contrarian view of the wisdom of the body as it relates to dietary self-wisdom. *Psychological Review*, **98**, 218–23.

Galef, B.G. & Henderson, P.W. (1972). Mother's milk: A determinant of the feeding preferences of weaning rat pups. *Journal of Comparative and Physiological Psychology*, **78**, 213–19.

Galef, B.G. & Sherry, D.F. (1973). Mother's milk: a medium for transmission of cues reflecting the flavour of mother's diet. *Journal of Comparative and Physiological Psychology*, **83**, 374–8.

Garr, J.L. & Stunkard, A.J. (1974). Taste aversions in man. *American Journal of Psychiatry*, **131**, 1204–7.

Greene, L.S., Desor, J.A. & Maller, O. (1975). Heredity and experience: Their relative importance in the development of taste preference in man. *Journal of Comparative and Physiological Psychology*, **89**, 279–84.

Harris, G. & Booth, D.A. (1985). Sodium preference in food and previous dietary experience in 6-month-old infants. *IRCS Medical Science*, **13**, 1177–8.

Hill, D.L. & Przekop, P.R., Jr. (1988). Influences of dietary sodium on functional taste receptor development: a sensitive period. *Science*, **241**, 1826–8.

Hill, D.L., Mistretta, C.M. & Bradley, R.M. (1986). Effects of dietary NaCl deprivation during early development on behavioural and neurophysiological taste responses. *Behavioral Neuroscience*, **100**, 390–8.

Johnson, S.L., McPhee, L. & Birch, L.L. (1991). Conditioned preferences: children prefer flavors associated with high dietary fat. *Physiology & Behavior*, **50**, 1245–51.

Kern, D.L., McPhee, L., Fisher, J., Johnson, S. & Birch, L.L. (1993). The postingestive consequences of fat condition preferences for flavors associated with high dietary fat. *Physiology & Behavior*, **54**, 71–6.

Krondl, M., Coleman, P., Wade, J. & Milner, J. (1983). A twin study examining the genetic influence on food selection. *Human Nutrition: Applied Nutrition*, **37A**, 189–98.

McBurney, D.H. & Gent, J.F. (1979). On the nature of taste qualities. *Psychological Bulletin*, **86**, 151–67.

Mattes, R.D. (1991). Learned food aversions: a family study. *Physiology and Behavior*, **50**, 499–504.

Mela, D.J. (1990). The basis of dietary fat preferences. *Trends in Food Science and Technology*, **1**, 71–3.

Mela, D.J. (1992). The perception and acceptance of dietary fat: what, who, why? *British Nutrition Foundation Nutrition Bulletin*, **17**, 74–86.

Mennella, J.A. & Beauchamp, G.K. (1991a). Maternal diet alters the sensory qualities of human milk and the nursling's behavior. *Journal of Paediatrics*, **88**, 737–44.

Mennella, J.A. & Beauchamp, G.K. (1991b). The transfer of alcohol to human milk. *New England Journal of medicine*, **325**, 981–5.

Mennella, J.A. & Beauchamp, G.K. (1993a). The effects of repeated exposure to garlic-flavoured milk on the nursling's behaviour. *Pediatric Research*, **34**, 805–8.

Mennella, J.A. & Beauchamp, G.K. (1993b). Beer, breast feeding, and folklore. *Developmental Psychobiology*, **26**, 459–66.

Moskowitz, H.W., Kumaraiah, V., Sharma, K.N., Jacobs, H.L. & Sharma, S.D. (1975). Cross-cultural differences in simple taste preferences. *Science*, **190**,

1217–18.
Pangborn, R. (1981). Individuality in responses to sensory stimuli. In *Criteria of Food Acceptance: How Man Chooses What He Eats*, ed. J. Solms and R.L. Hall, pp. 177–219. Zurich: Forster Verlag.
Pangborn, R.M., Guinard, J.-X. & Davis, R.G. (1988). Regional food preferences. *Food Quality and Preference*, 1, 11–19.
Pliner, P. (1982). The effects of mere exposure on liking for edible substances. *Appetite*, 3, 283–90.
Pliner, P. & Pelchat, M.L. (1986). Similarities in food preferences between children and their siblings. *Appetite*, 7, 333–42.
Poindron, P., Nowak, R., Lévy, F., Porter, R.H. & Schaal, B. (1993). Development of exclusive mother–young bonding in sheep and goats. *Oxford Review of Reproductive Biology*, 15, 311–64.
Porter, R.H., Makin, J.W., Davis, L.B. & Christensen, K.M. (1991). An assessment of the salient olfactory environment of formula-fed infants. *Physiology and Behavior*, 50, 907–11.
Prescott, J., Laing, D., Bell, G., Yoshida, M., Gillmore, R., Allen, S., Yamazaki, K. & Ishii, R. (1992). Hedonic responses to taste solutions: a cross-cultural study of Japanese and Australians. *Chemical Senses*, 17, 801–9.
Ramirez, I. (1990). Why do sugars taste good? *Neuroscience and Biobehavioral Reviews*, 14, 125–34.
Ray, J.W. & Klesges, R.C. (1993). Influences on the eating behaviour of children. *Annals of the New York Academy of Sciences*, 699, 57–69.
Rogers, P.J. & Lloyd, H.M. (1994). Nutrition and mental performance. *Proceedings of the Nutrition Society*, 54, 443–56.
Rogers, P.J. & Richardson, N.J. (1993). Why do we like drinks that contain caffeine? *Trends in Food Science and Technology*, 4, 108–11.
Rossi, J. (1980). La theorie du conditionnement alimentaire et son application lors du sevrage des porcelets. *Fortschritte Tierphysiol Tierernahr*, Pt. 2, 133–40.
Rozin, P. (1989). The role of learning in the acquisition of food preferences by humans. In: *Handbook of the Psychophysiology of Human Eating*, ed. R. Shepherd, pp. 205–27. Chichester UK: Wiley.
Rozin, P. (1991a). Family resemblance in food and other domains: the family paradox and the role of parental congruence. *Appetite*, 16, 93–102.
Rozin, P. (1991b). Getting to like the burn of chili pepper. Biological, psychological, and cultural perspectives. In *Chemical Senses. Volume 2: Irritation*, ed. B.G. Green, J.R. Mason and M.R. Kare, pp. 231–69. New York: Marcel Dekker, Inc.
Rozin, P. & Fallon, A.E. (1981). The acquisition of likes and dislikes for foods. In *Criteria of food acceptance: How man chooses what he eats*, ed. J. Solms and R.L. Hall, pp. 35–48. Zurich: Forster Verlag.
Rozin, P., Gruss, L. & Berk, G. (1979). Reversal of innate aversions: attempts to induce a preference for chili peppers in rats. *Journal of Comparative and Physiological Psychology*, 93, 1001–14.
Rozin, P. & Millman, L. (1987). Family environment, not heredity, accounts for family resemblances in food preferences and attitudes: a twin study. *Appetite*, 8, 125–34.
Rozin, P.N. & Schulkin, J. (1990). Food selection. In *Handbook of Behavioral Neurobiology*, ed. E.M. Stricker, pp. 297–328. New York: Plenum.

Schaal, B. & Orgeur, P. (1992). Olfaction *in utero*: can the rodent model be generalized? *Quarterly Journal of Experimental Psychology*, **44B**, 245–78.

Schleidt, M., Neumann, P. & Morishita, H. (1988). Pleasure and disgust: Memories and associations of pleasant and unpleasant odours in Germany and Japan. *Chemical Senses*, **13**, 279–93.

Sclafani, A. (1990). Nutritionally based learned flavor preferences in rats. In *Taste, Experience, and Feeding*, ed. E.D. Capaldi and T.L. Powley, pp. 139–56. Washington, DC: American Psychological Association.

Shepherd, R., ed. (1989). *Handbook of the Psychophysiology of Human Eating*. Chichester, UK: Wiley.

Story, M. & Brown, J.E. (1987). Do young children instinctively know what to eat? *New England Journal of Medicine*, **316**, 103–6.

Sullivan, S.A. & Birch L.L. (1990). Pass the sugar, pass the salt: Experience dictates preference. *Developmental Psychobiology*, **26**, 546–51.

Sullivan, S.A. & Birch, L.L. (1994). Infant dietary experience and acceptance of solid foods. *Pediatrics*, **93**, 271–7.

Sullivan, R.M., Taborsky-Barba, S., Mendoza, R., Itano, A., Leon, M., Cotman, C.W., Payne, T.F. & Lott, I. (1991). Olfactory classical conditioning in neonates. *Pediatrics*, **87**, 511–18.

Vogt, M.B. & Hill, D.L. (1993). Enduring alterations in neurophysiological taste responses after early dietary sodium deprivation. *Journal of Neurophysiology*, **69**, 832–41.

Wallace, J.P., Inbar, G. & Ernsthausen, K. (1992). Infant acceptance of postexercise breast milk. *Pediatrics*, **82**, 1245–7.

10 *The origins of coronary heart disease in early life*

D.J.P. BARKER

Introduction

Recent research has shown that babies who were small at birth and during infancy will be at increased risk of developing coronary heart disease, stroke, diabetes or hypertension during adult life. That a person's lifespan may be determined before birth is well known. Genetically determined diseases such as Huntington's chorea illustrate how a long period of normal development and adult life can be prematurely brought to an end by the action of inherited defects. What is new is the realization that the way in which gene expression may be permanently changed by the nutrient environment in early life can also affect the human lifespan.

There are three reasons why this new field of research has developed. First, the current explanation of coronary heart disease, a 'destructive' model in which inappropriate adult lifestyles hasten ageing processes, fails to account for either the time trends of the disease, or its geography, or why one person gets the disease and another does not. Secondly, the search for alternative explanations led to a strong geographical clue that the role of fetal life in the genesis of coronary heart disease might be much greater than had been thought (Barker & Osmond, 1986). Thirdly, animal experiments show that changes in nutrition in early life permanently change the growth and form of the body, together with a range of its structures and functions (McCance & Widdowson, 1974).

Studies in animals

The substantial body of evidence on the plasticity of the fetus, its ability to adapt to undernutrition, and the permanent effects of these adaptations, derive from animal experiments carried out by Widdowson and others. These studies allow us to predict two things about the human fetus. First, lack of nutrients or oxygen will cause persisting changes, which include

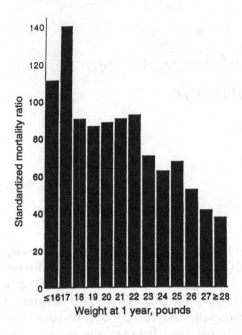

Figure 10.1. Standardized mortality ratios for coronary heart disease in 8175 men according to their weight at one year of age.

altered metabolism, including glucose and lipid metabolism, altered blood pressure and altered 'settings' of hormonal axes, enzymes and cell receptors. Secondly, the long-term effects of undernutrition depend on the stage at which it occurs. Tissues and systems tend to be vulnerable to programming during phases of rapid cell replication, and different tissues undergo these 'sensitive' phases of development at different times.

Small size at birth and in infancy

It has been possible to explore the links between growth *in utero* and later coronary heart disease because a search of the archives in Britain revealed a number of collections of birth records of men and women born 50 years and more ago. Figure 10.1 shows findings in a group of 8175 men born in the county of Hertfordshire before 1930. Their weight at one year of age strongly predicted their subsequent death rates from coronary heart disease (Osmond *et al.*, 1993). Death rates fell steeply between those who were small and those who were large at one year. There were similar trends in coronary heart disease with birthweight in men and women. A study in Sheffield showed that the small babies with high coronary death rates were

Table 10.1. *Mean systolic pressure (mmHg) in men and women aged 64–71 years according to birthweight*

Birthweight pounds (kg)	Men	Women
−5.5 (2.50)	171 (18)	169 (9)
−6.5 (2.95)	168 (53)	165 (33)
−7.5 (3.41)	168 (144)	160 (68)
−8.5 (3.86)	165 (111)	163 (48)
>8.5 (3.86)	163 (92)	155 (26)
Total	166 (418)	161 (184)
Standard deviation	24	26

Figures in brackets are numbers of subjects.

small in relation to the duration of gestation, rather than small because they were prematurely born.

These findings pose the question of what are the processes which link reduced early growth with adult disease. From examining samples of men and women who still live in Hertfordshire, and Sheffield, and in Preston, we now know that babies who were small have, as adults, raised blood pressure, raised serum cholesterol and plasma fibrinogen concentrations and impaired glucose tolerance (Barker *et al.*, 1993), the main risk factors for coronary heart disease. Table 10.1 shows the mean systolic pressures of men and women aged 64–71 years. Systolic pressure falls progressively between those who were small at birth and those who were large. The relation between birthweight and blood pressure has now been demonstrated in 21 studies of children and adults, and there is a secure base for saying that impaired fetal growth is strongly linked to blood pressure at all ages except during adolescence, when the tracking of blood pressure levels which begins in early childhood is perturbed by the adolescent growth spurt. Differences in blood pressure associated with birthweight are small in childhood but are magnified throughout life. This suggests that there may be amplification as well as initiation processes. We do not know what initiates high blood pressure in intrauterine life, but there are interesting clues including work which has pointed to the possible role of cortisol (Benediktsson *et al.*, 1993).

Table 10.2 shows the prevalence of non-insulin-dependent diabetes and impaired glucose tolerance according to birthweight in a group of men in Hertfordshire (Hales *et al.*, 1991). The prevalence falls sharply between men who were small at birth and men who were large. There are similar

Table 10.2. *Prevalence of non-insulin-dependent diabetes and impaired glucose tolerance in men aged 59–70 years*

Birthweight pounds (kg)	Number of men	% with impaired glucose tolerance or diabetes	Odds ratio adjusted for body mass index (95% confidence interval)
≤5.5(2.50)	20	40	6.6(1.5 to 28)
−6.5(2.95)	47	34	4.8 (1.3 to 17)
−7.5(3.41)	104	31	4.6(1.4 to 16)
−8.5(3.86)	117	22	2.6(0.8 to 8.9)
−9.5(4.31)	54	13	1.4(0.3 to 5.6)
>9.5(4.31)	28	14	1.0
Total	370	25	

findings in women.

Body proportions at birth

Studies of men and women who were small at birth have shown that they are resistant to insulin. The occurrence of insulin-resistance in adults is characterized in Syndrome X, in which diabetes, hypertension and raised plasma triglyceride concentrations coincide in the same person. Table 10.3 shows that, allowing for current body mass, the relative risk of having Syndrome X among people who were 6.5 pounds (2.95 kg) or less at birth is around ten times higher than among people who were more than 9.5 pounds (4.31 kg). For comparison, the risk of coronary heart disease among smokers compared with non-smokers is around 2. The insulin resistance syndrome is associated not only with low birthweight but with thinness at birth, as measured by a low ponderal index (birthweight/length^3). Babies who are thin at birth lack muscle as well as fat and muscle in adult life is the peripheral site of insulin action. Insulin tolerance tests on men and women aged 50 years confirm that those who were thin at birth are less sensitive to insulin.

Raised blood pressure in adult life is associated not only with thinness at birth but also with short body length in relation to head size. Short babies are thought to have encountered undernutrition in late gestation and to have sustained brain growth at the expense of the trunk, including the abdominal viscera. Table 10.4 shows mean serum cholesterol concentrations in a group of men and women aged 50 years according to abdominal circumference at birth. The concentrations of total and LDL cholesterol fall between people who had small and large abdominal circumferences

Table 10.3. *Prevalence of Syndrome X (type 2 diabetes, hypertension and hyperlipidaemia) in men according to birthweight*

Birthweight pounds (kg)	Total number of men	% with Syndrome X	Odds ratio adjusted for body mass index (95% confidence interval)
≤5.5(2.50)	20	30	18 (2.6 to 118)
−6.5(2.95)	54	19	8.4(1.5 to 49)
−7.5 (3.41)	114	17	8.5(1.5 to 46)
−8.5(3.86)	123	12	4.9(0.9 to 27)
−9.5(4.31)	64	6	2.2(0.3 to 14)
>9.5(4.31)	32	6	1.0
Total	407	14	

Table 10.4. *Mean serum lipid concentrations according to abdominal circumference at birth in men and women aged 50–53 years*

Abdominal circumference (inches)	Number of people	Total cholesterol (mmol/l)	Low density lipoprotein cholesterol (mmol/l)
−11.5	53	6.7	4.5
−12.0	43	6.9	4.6
−12.5	31	6.8	4.4
−13.0	45	6.2	4.0
>13.0	45	6.1	4.0
Total	217	6.5	4.3

(Barker *et al.*, 1993). Abdominal circumference reflects liver size, the liver being disproportionately large in the fetus. An inference from Table 10.4 is that babies who have impaired liver development permanently re-set their cholesterol metabolism. Reduced abdominal circumference at birth is also associated with raised plasma concentrations of fibrinogen, another strong predictor of coronary heart disease. The differences in serum cholesterol and plasma fibrinogen concentrations associated with the range of abdominal circumference at birth and infant growth are large, equivalent to at least 30% differences in risk of coronary heart disease.

Summary of programming

This brief review of what is known in animals and man allows a number of conclusions.

1. Restriction of nutrients or oxygen *in utero* leaves permanent marks on the physiology and structure of the body. As an example, Table 10.5 shows

Table 10.5. *Effects of fetal exposure to maternal low protein diets on sys-tolic blood pressure in adult rats*

Dietary protein percent by weight	Number	Mean (SD) systolic blood pressure 9 weeks after birth (mm Hg)
18	15	137 ($\pm$4)
12	13	152 ($\pm$3)
9	13	153 ($\pm$3)
6	11	159 ($\pm$4)

the blood pressures of the offspring of four groups of pregnant rats given varying amounts of dietary protein (Langley & Jackson, 1994). The offspring of rats who had lower protein diets had raised blood pressures nine weeks after birth and this persisted through adult life.

2. Experiments on animals have established that undernutrition at different times in early life has different effects. Undernutrition in early gestation leads to proportionate loss of body size as in the proportionally small newborn human baby. In late gestation undernutrition leads to disproportionate growth, as in the thin or short human baby. Dispropor-tionate growth rather than small size seems to hold a key to the origins of coronary heart disease. Undernutrition can effect profound changes in the relative size of the body's organs without any major change in overall body size (McCance & Widdowson, 1974).

3. The rapidly growing baby is more vulnerable to undernutrition. When rickets was common 70 years ago, it was not small babies who got the disease but larger, more rapidly growing ones. Slow growth protects against undernutrition. In some countries such as China, where in-trauterine growth retardation is widespread, coronary heart disease is rare. Growth retardation in China seems to lead to downregulation of growth in early gestation, which could protect the fetus from the effects of undernutri-tion later in gestation, and from the development of the disproportion which is associated with coronary heart disease.

4. Fetal undernutrition, which programmes the body, itself results from inadequate maternal intake of food, or inadequate transport or transfer of nutrients. Studies of the birthweights of families show a strong correlation between the birthweights of people related through their mothers, but not between the birthweights of people related only through their fathers. This and other findings suggest that fetal growth is not predominantly controlled by the fetal genome but by the supply of nutrients and oxygen from the mother (Barker, 1994). In 1944, for a period of seven months, there was an embargo on food supplies to the population of western

Table 10.6. *Mean systolic blood pressure (mm Hg) of men and women aged 46 to 54 according to placental weight and birthweight*

Birthweight (pounds)	Placental weight (pounds)				
	1.0	−1.25	−1.5	>1.5	All
<5.5	152(26)	154(13)	153(5)	206(1)	154(45)
−6.5	147(16)	151(54)	150(28)	166(8)	151(106)
−7.5	144(20)	148(77)	145(45)	160(27)	149(169)
>7.5	133(6)	148(27)	147(42)	154(54)	149(129)
All	147(68)	149(171)	147(120)	157(90)	150(449)

Number of subjects in parentheses.

Holland. People starved. A generation of babies were conceived or born during famine and we now know something about what happened to them as adults (Lumey, 1992). Girls who were conceived in the famine but born after liberation by the allies, had normal birthweight and grew up to be normal women, but their babies, when they were born, were small. The ability of these women to deliver nutrients to their babies had, it seems, been impaired by their own fetal experience. This observation illustrates how fetal nutrition depends not only on what the mother eats during pregnancy but on her physiological and metabolic competence, established during her early life, as well as her nutrient stores before pregnancy.

Table 10.6 shows another aspect of the complex links between maternal and fetal nutrition. The mean systolic pressures of a group of men and women are arranged by four groups of birthweight and four groups of placental weight. As expected from previous findings, those who were of heavier birthweight had lower blood pressure. But, unexpectedly, at any birthweight, men and women who had had larger placentas had higher blood pressure. From studies in animals we know that placental enlargement is an adaptation to lack of nutrients, including oxygen; and in humans three kinds of baby are known to have disproportionately large placentae. They are the offspring of mothers who were anaemic in pregnancy, who exercised during pregnancy or who live at high altitude (Barker, 1994). The fetus it seems attempts to overcome the deficiency in supply of nutrients or oxygen to it by increasing the area of its attachment to the mother. A high ratio of placental weight to birthweight is linked to cardiovascular disease, impaired glucose tolerance and raised plasma fibrinogen concentrations in later life as well as to hypertension. The placenta seems to play an important role in programming the baby.

162   D.J.P. Barker

New model of coronary heart disease

A new model for the causation of coronary heart disease is emerging
(Barker, 1994). Under the old model an inappropriate lifestyle, including
cigarette smoking and lack of exercise, leads to accelerated destruction of
the body in middle and late life, including the more rapid development of
atheroma, raised blood pressure, and the development of insulin resistance.
Under the new model, coronary heart disease results not primarily from
external forces but from the body's internal environment, homeostatic
settings of enzyme activity, cell receptors, and hormone feedback, which
are established in response to undernutrition *in utero* and lead eventually to
premature death.

References

Barker, D.J.P. (1994). *Mothers, Babies and Disease in Later Life*. London: British
    Medical Journal Publications.
Barker, D.J.P. & Osmond, C. (1986). Infant mortality, childhood nutrition, and
    ischaemic heart disease in England and Wales. *Lancet*, i, 1977–81.
Barker, D.J.P., Gluckman, P.D., Godfrey, K.M., Harding, J.E., Owens, J.A. &
    Robinson, J.S. (1993). Fetal nutrition and cardiovascular disease in adult life.
    Lancet, 341, 938–1.
Barker, D.J.P., Martyn, C.N., Osmond, C., Hales, C.N., Jespersen, S. & Fall,
    C.H.D. (1993). Growth in utero and serum cholesterol concentrations in adult
    life. *British Medical Journal*, 307, 1524–7.
Benediktsson, R., Lindsay, R.S., Noble, J., Seckl, J.R. & Edwards, C.R.W. (1993).
    Glucocorticoid exposure *in utero*: new model for adult hypertension. *Lancet*,
    341, 339–41.
Hales, C.N., Barker, D.J.P., Clark, P.M.S., Cox, L.J., Fall, C., Osmond, C. &
    Winter P.D. (1991). Fetal and infant growth and impaired glucose tolerance at
    age 64 years. *British Medical Journal*, 303, 1019–22.
Langley, S.C. & Jackson, A.A. (1994). Increased systolic pressure in adult rats
    induced by fetal exposure to maternal low protein diets. *Clinical Science*, 86,
    217–22.
Lumey, L.H. (1992). Decreased birthweights in infants after maternal *in utero*
    exposure to the Dutch famine of 1944–45. *Paediatric and Perinatal Epidemiol-
    ogy*, 6, 240–53.
McCance, R.A. & Widdowson, E.M. (1974). The determinants of growth and
    form. *Proceedings of the Royal Society of London (Series Biology)*, 185, 1–17.
Osmond, C., Barker, D.J.P., Winter, P.D., Fall, C.H.D. & Simmonds, S.J. (1993).
    Early growth and death from cardiovascular disease in women. *British
    Medical Journal*, 307, 1519–24.

11 Early life stresses and adult health: insights from dental enamel development

ALAN H. GOODMAN

Introduction

Because early life is often a stage of increased physiological stress, morbidity, and mortality, this period is a justified focus of biomedical research and public health intervention efforts (Mosley, 1983; Preston, 1980; Wood, 1983). Indeed, a widely accepted public health goal is to increase survival to age five (Mosley, 1983). In contrast, an up-to-now infrequently considered question concerns the consequences in adulthood of survived, early life stress. What happens to the survivors of undernutrition and other early perturbations? Do these individuals go on to lead healthy and high functioning lives as adults, or do they suffer illnesses more often and die younger? What biocultural processes might link early life events and stresses to later health and functioning?

Although questions pertaining to the long-term consequences of early stress are both fascinating and important, they have not been the subject of sustained and systematic investigation, most likely because of epidemiological difficulties in linking present events to events in the distant past. Thus, the main purpose of this paper is to critically evaluate potential hard tissue 'windows' or memories of early life stress. If these hard tissue measures are available in adulthood as reliable, valid, and unaltered epidemiological measures of early events, they may provide a unique means for linking early events to adult conditions. Finally, whereas this chapter focuses on applications from past populations, in fact mainly 'skeletal series', the methods are not limited to past or skeletal populations. Indeed, a main conclusion is that studies of hard tissue memories of past physiological perturbations should be undertaken with living individuals and groups, in situations in which the results can be contextualized and processes better understood. In this chapter: (i) common epidemiological

163

difficulties in linking early stress and adult health are briefly reviewed; (ii) various bone and tooth bioassays that might be used as measures of early life stress are compared; (iii) enamel, and specifically linear enamel hypoplasias (LEH), a developmental enamel defect, are highlighted as they seem to have the greatest potential for these types of epidemiological concerns; (iv) two examples are reviewed on the use of LEH in linking early and later health (one from Dickson Mounds, an archaeological population, and the other from Hammon-Todd, a historical skeletal series); and (v) some conclusions and implications are suggested for better determining the processes by which early and late events may be linked.

Linking early events and adult health

A prominent assumption underlying much of the literature on development and ageing is that stressful events occurring early in development may influence adult health. Freud, for example, believed that adult phobias were ultimately caused by traumas occurring during well-defined psycho-developmental stages. Whereas it is ambiguous in many other areas, psychoanalytical theory is clear in predicting that the type and severity of mental illness is a function of the severity and timing of traumas during development.

Aetiological links between childhood stress and adult health are also prominent in the physical health literature. All but a few theories of ageing leave a place for the influence of early events. A variety of 'wear and tear' theories predict that repeated or chronic stress will predispose to an increased rate of loss of functional abilities, more illness, and early mortality. Although common, these assumptions and theories are hard to evaluate. Controlled animal experiments have provided a variety of leads (Herrenkohl, 1979), but the applicability of these studies to humans is uncertain.

Studies of the long-term consequences of early stress in humans may be divided into a few fundamental types. The most robust experimental design involves prospective studies in which individuals have been followed throughout life (Wadsworth, 1986). Unfortunately, there are few such studies: they are costly, difficult to conduct and maintain, and, by definition, they take a long time. A more common type of study is based on retrospective reporting of a stressful event during early life. Whereas these case control designs can be useful, they all suffer from the vagaries of individual memories. This recall error may be eliminated in studies of historically documented stressful events such as the Dutch Famine of 1944–45 (Stein *et al.*, 1975) and the experience of living in a concentration camp (Schmolling, 1984). These unique 'natural' experiments are likely to continue to provide key insights into the long-term consequences of early stresses.

Other researchers have avoided the vagaries of memory by recovering records of health and development from the prenatal and neonatal period, and linking it to adult morbidity and mortality. This might be done on the level of the population or the individual. For example, in an ecological design, Barker and colleagues have frequently linked known historical socioeconomic and health conditions to geographic variation in health in more recent times (Barker & Osmond, 1986a,b, 1987), or have followed individuals with early health/nutritional status records, such as birth-weights, and contrast health experiences in adulthood in relationship to the records from early life (see Barker et al., 1990, 1992).

In a sense, the bioassays that are reviewed below are employed in this last way. The only difference is that a currently measurable biological characteristic replaces a written record. This is exactly what is done when adult height is employed as a measure of early life nutritional status (D'Avanzo, La Vecchia & Negri, 1994; Marmot, Shipley & Rose, 1984; Waaler, 1984). The obvious advantage of this method is that one need not depend on the availability of prior health records; early life nutrition and health status is measured at the same time that current health is assessed.

In summary, the tool kit for studying the relationship between early stress and adult health, and making sense of these relationships, is rather sparse. Methods generally trade off reduced temporal and financial costs for epidemiological soundness. Moreover, it is harder still to interpret processes that link past and present. How, for example, does one control for the socioeconomic conditions that may effect exposure to early, intermediate, and late stressors? None the less, these are interesting and important questions and with some creativity they may be studied empirically. The aim of this chapter is to suggest some further research tools to bring to light these hidden relationships.

Skeletal memories of stress

In studying health in past populations; paleoepidemiologists have only bones and teeth as guides. Although limited, a variety of skeletal measures of early life perturbation and disease have been proposed and utilized, and furthermore, the structures upon which these measures are based stabilize and may be discerned in adult skeletons (Goodman & Armelagos, 1989; Martin, Goodman & Armelagos, 1985). Common examples of these measures of childhood stress or ill-health seen in adult skeletons include porotic hyperostosis, bone infection (periostitis), and the developmental indicators of Harris lines, skull base height, vertebral neural canal (VNC) diameters, long bone lengths and widths, and developmental enamel defects (specifically linear enamel hypoplasia) (Goodman & Armelagos, 1989; Martin *et al.*, 1985).

Porotic hyperostosis, usually interpreted to be a sign of iron-related anaemia (Martin *et al.*, 1985), and periostitis, a sign of generalized bone infection, are bone pathologies that can be observed in adulthood, and with some assurance be traced to an infant–childhood origin. Mittler and Van Gerven (1994) have recently shown that cribra orbitalia, a type of porotic hyperostosis, is linked dramatically to decreased longevity in an archaeological population from Sudanese Nubia. Unfortunately, it is not always easy to estimate the age at which a lesion developed, since bone turnover and remodelling does occur thus obscuring some lesions, and this method has little applicability to living population research because these bone pathologies are greatly obscured by skin and soft tissue layers.

The use of VNC diameters (or stenosis), skull base height, and long bone lengths and widths is based, along with the study of adult height, on the common proposition that adult size variation is attributable fundamentally to growth performance during a critical, early period. Clark *et al.* (1986), for example, argue that the VNC dimensions are stabilized by about four years of age, and thus reflect early health conditions, in fact, mirroring strongly the conditions of neural and lymphatic organ development. They have linked VNC diameters to decreased longevity in an archaeological population and have suggested its use via modern xerography as a way to predict adult morbidity and mortality in living populations. Whereas this method is promising in theory, research has not established unambiguously the time of development of VNC dimensions, the potential to change VNC dimensions after childhood, or the specificity and sensitivity of the VNC to common stressors. The same unknowns and limitations apply to skull base height and other osseous measurements. Finally, whereas a great deal is known about the sensitivity and specificity of height to undernutrition and disease stresses, there is still much debate as to the ubiquity and degree to which catch-up in height occurs (see, for contrasting views, Martorell *et al.*, 1990, and Golden, 1994), and by extension the validity of adult height as a measure of conditions during infancy and childhood.

Harris lines, or growth arrest lines, are transverse lines of increased radiopacity seen on X-rays, particularly of tibia and other long bones (Martin *et al.*, 1985). The position of the lines reflect the size of the bone shaft at the time of formation, and from this datum one can extrapolate developmental age at time of line formation (Martin *et al.*, 1985; Magennis, 1990). It was also once assumed that these lines reflected periods of relatively acute stress. If these propositions are true, and if lines infrequently remodel and resorb, then a chronological record of physiological perturbations might be obtained in an adult by locating lines on long bone radiographs. Unfortunately, Harris lines do resorb, and even more unfortunately, Harris lines do not appear to be valid measures of stress (Martin *et al.*, 1985). Magennis (1990), for example, finds little association with childhood disease, and shows that Harris lines are more (not less) frequent during periods of rapid growth in length. Thus, there are essential reasons to suggest that bone measurements, including heights, are not the clearest possible windows into past conditions. None the less, a few possibilities do exist and they should be further evaluated.

Linear enamel hypoplasias

Another method of chronologically assessing early life stress is found in the record of disturbed enamel quality and quantity. Two unambiguous and relatively unique advantages of enamel as a tissue are that it is easy to visually inspect and it does not remodel. Enamel is essentially inert once formed and calcified; once enamel matures, its structure is unalterable by internal biological events. Additionally, since enamel is secreted in a regular and ring-like fashion, the tooth's enamel crown provides a permanent chronological record of metabolic disruptions which occurred during its time of development (Kreshover, 1960; Sarnat & Schour, 1941; Via & Churchill, 1959).

Linear enamel hypoplasias (LEHs) are a class of developmental enamel defects (DDE). They are seen visually as circumferential areas of decreased enamel thickness (Goodman & Rose, 1990, 1991). They are the direct result of a disruption in ameloblastic matrix secretion (Goodman & Rose, 1990), a conclusion that is over a half century old (Sarnat & Schour, 1941). Enamel hypoplastic defects may be due to hereditary conditions, localized trauma, or systemic disruption (stress) (Pindborg, 1970, 1982). Hypoplastic defects which are due to systemic disruptions (chronological enamel hypoplasias) may be distinguished from those which are the result of other factors (Goodman & Rose, 1990; Yaeger, 1980). Systemic disruptions are likely to affect more than one tooth, and the location of the defect on these teeth will reflect the relative completeness of crown development at the time of the stress (Sarnat & Schour, 1941; Yaeger, 1980; also see Figure 11.1).

The precise aetiology of enamel hypoplasias is unknown. A long list of environmental, nutritional, physiological and hormonal conditions have been aetiologically linked to LEH formation (Goodman & Rose, 1990). Work with contemporary children in Mexico, for example, shows that LEH is more common in families with less material wealth and is predictive of decreased growth status (height-for-age and weight-for-age) (Goodman *et al.*, 1992). Another study, comparing supplemented and non-supple-

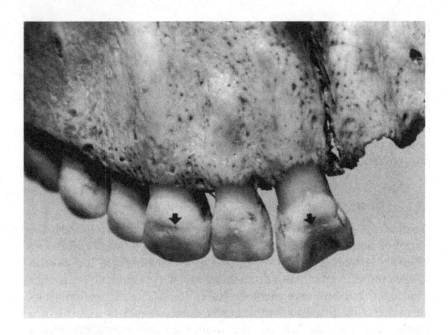

Figure 11.1. Chronologic enamel hypoplasias (stress-hypoplasias) on right maximally central incisor and canine (arrow). Both of these shallow hypoplastic bands developed around 3.5 year developmental age, based on the degree of crown completion (enamel apposition begins at the occlusal tips which are slightly worn). The common estimated age at formation suggests that these defects were the result of the same systemic physiological perturbation (stress).

mented children, found that enamel defects are nearly twice as common in non-supplemented children (Goodman, Martinez & Chavez, 1991). Thus, a working 'threshold' model has been proposed in which enamel defects formation commences when a physiological threshold is passed, and enamel matrix formation stops (Figure 11.2). The threshold is influenced by nutrient intake, concurrent morbidity, and 'unknown aetiological factors'.

The aetiology of LEH may be seen as very similar to the aetiology of linear growth retardation. The differences are that enamel disruption is more of a discontinuous, threshold variable. Another difference, of potential importance in considering long-term consequences, is that enamel is an epithelial tissue and its disruption may be concurrent with the disruption in development of other epithelial tissues. For example, Jaffe *et al.* (1985) have shown that individuals with idiopathic brain disorders are much more likely to also have disrupted enamel development than normal

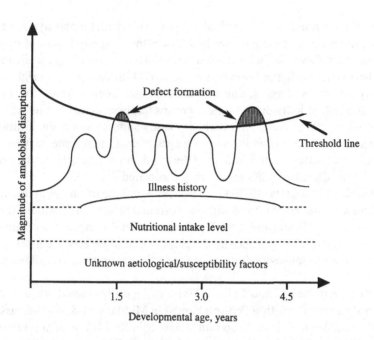

Figure 11.2. Threshold model for the formation of a linear enamel hypoplasia (LEH). Enamel apposition and LEH formation occurs when ameloblasts temporarily cease secretion of enamel matrix. This growth stoppage has been associated with a wide variety of stressors. It is suggested that there is a threshold at which aneloblasts will cease function. This threshold may be reached by a combination of unknown etiological factors, chronic or acute undernutrition and acute morbidity.

controls. This association points to a common physiological perturbation during early gestation that affected both epithelial tissues.

Prior studies of LEH and adult mortality

A few sporadic studies have examined the association between LEH and future mortality. The earliest of these was a study of living Chinese (Anderson & Stevenson, 1930). These researchers found that enamel defects were far less common in older cohorts. They attributed this finding to a selecting out of individuals who were stressed early in life. Unfortunately, it is not clear what specific type of defect they were observing (it is called 'mottled enamel').

The first palaeontological/archaeological data suggesting decreased longevity with childhood LEH comes from the late 1970s. White (1978) assessed hypoplasias on permanent maxillary first molars from South African Plio-Pleistocene Australopithecines (ca. 1.5–3.0 million years before present). He noted that individuals with maxillary first molar hypoplasias from the Swartkrans site ($n = 6$) had 'lower-than-expected' ages at death. These individuals died between four and thirteen years of age, while individuals with non-hypoplastic first molars ($n = 110$) died between eight and thirty-one years of age. Using White's estimated ages-at-death, the mean age at death of individuals in these groups has been calculated as 7.8 and 19.6 years, respectively. Although this study suffers from a small sample size and lack of precision in assigning ages of death to fragmentary palaeontological materials, the data none the less demonstrate a dramatic, nearly 12-year decrease in life expectancy associated with LEHs.

Cook and Buikstra (1979) similarly compared the mean age at death of infants and children with and without postnatal defects on deciduous tooth crowns from Middle and Late Woodland skeletal samples from Illinois. They conclude that postnatally developing dental defects are associated with decreased longevity during both the Middle and Late Woodland periods.

Rose, Armelagos, and Lallo (1978) histologically studied areas of disturbed enamel formation (Wilson bands) in Middle Woodland, Mississippian Acculturated Late Woodland, and Middle Mississippian samples from Illinois. They found that individuals with Wilson bands died at an earlier mean age at death in all samples. Overall, the average age at death of the 21 individuals with Wilson bands is 26.7 versus 42.1 years in the 66 individuals without Wilson bands. In other words, individuals with a record of early life stress (Wilson band) lived 15.4 years less than those without such a record.

Most recently, Duray (1994) has studied the association between age at death and developmental enamel defects at Libben, a large archaic cemetery in Ohio. Duray finds that individuals with developmental defects also die earlier, by about 5 to 6 years, depending on the type of defect, defect severity, and defect location.

These results vary by population studied and means of assessment of enamel developmental defects. However, of greatest note is that all studies, despite methodological and populational differences, suggest a measurable decrease in longevity associated with early life stress. This association, and the potential biocultural processes that might underlie it, are explored in greater detail in the following two case studies.

Dickson mounds: LEH, cultural change and life expectancy

Dickson Mounds is a multicomponent habitation-burial complex located near Lewiston, Illinois. The mounds are associated with three cultural horizons: Late Woodland (LW), Mississippian Acculturated Late Woodland (MALW), and Middle Mississippian (MM) (Harn, 1971, 1978, 1980). During the Late Woodland period (ca. AD 900–1050) the area was occupied by a relatively small (75–125) and semisedentary hunting and gathering population with seasonal camp sites and an economy directed toward the use of a broad spectrum of local fauna and flora. The MALW (ca. AD 1050–1175) is a transitional period, possible overlapping for some time with the LW, during which local populations began to come under the influence of Mississippian cultures further to the south in the American Bottom (Harn, 1978, 1986). During the MM period (ca. AD 1175–1300) the Mississippianization of local populations becomes complete with the culmination of trends toward extended and intensified trade networks, increased population density, size, and sedentism, and greater reliance on maize agriculture (Harn, 1978).

These changes have been associated with an increase in nutritional and infectious pathologies and a decrease in life expectancy (Goodman *et al.*, 1984*b*). Porotic hyperostosis, an indication of iron deficiency anaemia, is four times as prevalent among MM subadults as compared to LW subadults (64% to 16%) (Lallo, Armelagos & Mensforth, 1977). Periosteal infections in subadults increase from 27% in the LW to 81% in the MM (Lallo, Armelagos & Rose, 1978), and the frequency of enamel hypoplasias doubles in adults and adolescents (Goodman, Armelagos & Rose, 1980). Life expectancy is lower at all age intervals in the MM when compared to the combined LW and MALW samples (Moore, Swedlund & Armelagos, 1975). Long bone growth in length and circumference appears to be slower around 2–5 years of age and LEH and enamel histological defects are nearly twice as common in the MM as compared to the LW (Goodman *et al.*, 1980, 1984*b*).

Enamel hypoplasias were recorded on all permanent teeth except third molars in 111 adults and adolescents (50 males, 50 females, 11 adolescents of unknown sex). Hypoplasias are easily identified and were defined operationally as circumferential lines, bands, or pitting of decreased enamel thickness (Goodman *et al.*, 1980; 1984*a*) (Figure 11.1). The distance of the hypoplasia from the cemento-enamel junction was measured to one-tenth mm using a thin-tipped caliper. This distance measure was converted to the individual's dental age when the disruption occurred, based on the developmental standard of Massler and co-workers (1941).

Each half-year period between birth and 7.0 years was rated as either stress-positive or stress-negative depending on the identification of LEHs on areas of teeth developing during these time periods. By using this method, a chronology of stress by half-year periods was developed for each individual from birth to seven years of age (Goodman *et al.*, 1980, 1984*a*).

Table 11.1. *Comparison of mean ages at death for individuals by cultural horizon and number of hypoplasias-stress periods between 3.5 and 7.0 years developmental age*

	Sample size	Mean	S.D.	1-way ANOVA (F-ratio)	A priori contrasts (A vs. B)	(A vs. C)	T-values (A vs. B+C)
Late Woodland	20	33.0	11.5	.35	.55	—	.55
No hypoplasias (A)	11	31.6	10.4				
One hypoplasia (B)	9	34.7	13.0				
2–3 hypoplasias (C)	—	—	—				
MALW	45	33.3	13.4	1.44	1.22	1.53	1.69
No hypoplasias (A)	22	36.6	12.8				
One hypoplasia (B)	14	31.1	14.7				
2–3 hypoplasias (C)	9	28.6	11.7				
Middle Mississippian	46	31.6	11.2	6.52[c]	2.25[b]	3.50[d]	3.52[d]
No hypoplasias (A)	17	37.5	9.0				
One hypoplasia (B)	22	30.2	11.0				
2–3 hypoplasias (C)	7	21.8	8.7				
Total sample	111	32.5	12.1	4.99[c]	1.84[a]	3.04[c]	3.08[c]
No hypoplasias (A)	40	35.8	10.1				
One hypoplasia (B)	45	31.4	12.7				
2–3 hypoplasias (C)	16	25.6	10.8				

[a] 2-tailed $p \leq .10$.
[b] 2-tailed $p \leq .05$.
[c] 2-tailed $p \leq .01$.
[d] 2-tailed $p \leq .001$.

This study is based on the evidence for stress between 3.5 and 7.0 years of age. The extensive dental attrition characteristic of the Dickson series limited our ability to observe the enamel record of stress from birth to 3.5 years. Due mainly to occlusal surface attrition, many individuals have a series of undetermined periods starting at birth–0.5 years and extending as far as the 3.0–3.5 year period (Goodman *et al.*, 1984*a*). However, all individuals yielded a complete record of stress-hypoplasias for the seven half-year periods from 3.5 to 7.0 years.

All individuals in the LW sample have either one or no hypoplasias-stress periods between 3.5 and 7.0 years (Table 11.1). Individuals with one hypoplasia-stress period have a slightly greater mean age at death (34.7 years) than individuals with no hypoplasias-stress periods (31.6 years). This difference, however, is not statistically significant (Table 11.1; F-ratio = .35; Goodman & Armelagos, 1988).

This association between hypoplasias-stress periods and longevity is reversed during the MALW periods. The mean age at death of individuals without hypoplasias-stress periods is 36.6 years, or 5.5 years greater than those with one hypoplasia-stress period (31.1 years) and 8.0 years greater than those with two or more stress periods (Table 11.1, Figure 11.3).

This inverse association between stress periods and mean age at death is most pronounced during the Middle Mississippian. The mean age at death of individuals without hypoplasias-stress periods is 37.5 years, or 7.3 years longer than those with one hypoplasia-stress period and 15.7 years longer than those with two or more hypoplasias-stress periods (Figure 11.3). A one-way ANOVA, testing for the statistical significance of differences in ages at death among hypoplasia-stress period groups (Nie *et al.*, 1975), yielded an F-ratio of 6.52 (Table 11.1; p < .01).

The specificity of differences between hypoplasia groups was tested with a series of *a priori* contrasts (Table 11.1). These provide a comparison of the mean age at death in group A (no stress periods) with: 1) group B (one stress period), 2) group C (two or three stress periods), and 3) group B+C combined (one or more stress periods). For the MM group, all *a priori* contrasts yielded statistically significant results at the .05 level of confidence. The most significant differences are found in comparing individuals without any stress periods with those with one or more or two or more stress periods (t = 3.50 and 3.52; p < .001 and < .001).

Finally, there is a significant decrease in longevity with childhood stress periods in the total sample (Table 11.1; Figure 11.3). The overall mean age at death of individuals without hypoplasias-stress periods is 35.8 years; 4.4 years greater than for those with one hypoplasia-stress period (31.4 years) and 10.2 years greater than for those with two or more hypoplasia-stress periods (25.6 years).

174   *A.H. Goodman*

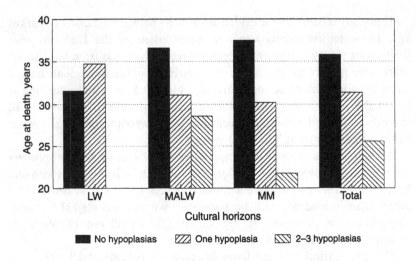

Figure 11.3. Mean ages at death of Dickson Mounds adolescents/adults by
number of hypoplasias-stress periods between 3.5 and 7.0 years developmental
age. LW = Late Woodland, MALW = Mississippian Acculturated Late
Woodland, MM = Middle Mississippian.

There are at least three processes which may account for the association
between childhood stress and decreased life expectancy. First, these data
may result from differential lifelong patterns of biological susceptibility to
physiological disruptions and their adverse effects. An increased susceptibility to stress may be causative of both an increased frequency of
childhood hypoplasias and an earlier age at death. That is, individuals who
are ill during childhood continue to fall ill as adults. This association is due
to a frailty or a 'weaker constitution', and the sum effect is earlier death.

Secondly, individuals who were exposed to, and survived, a period of
severe childhood stress may suffer a loss in ability to respond to other
stresses. In a sense, these individuals are 'biologically damaged' by the early
stress. The wear and tear of stresses during development may render them
less fit to respond to and survive subsequent stresses. For example,
suboptimal early nutrition has been proposed as a mechanism for later
immune disfunction (Chandra, 1975; Miler, 1982).

Thirdly, these data may result from differential lifelong patterns of
behavioural and culturally based exposure to stressors. An increased
lifelong potential for exposure to stressors may be causative of both an
increased frequency of childhood stress and earlier ages at death.

It is not possible to rule out any of these processes. All may contribute to
the associations which we have observed. However, the wide variation in

the degree of association between stress and longevity supports the view that the association is not solely a function of biological factors, since these samples appear to be genetically continuous (Cohen, 1974; Goodman *et al.*, 1984*a*). Furthermore, the greatest difference between stressed and non-stressed group mean ages at death occurs in the MM period. Since this is also the horizon in which status differences are likely to be greatest (Rothschild, 1979), these data suggest the importance of lifelong differences in social status and concomitant differential cultural buffering from stress. Unfortunately, it is difficult to assess cultural buffering in archaeological populations. We have tested to see if differences in type of grave offering, an indicator of status differences, might explain the association between stress and longevity (Goodman, Rothschild & Armelagos, 1983). Whereas individuals with no grave goods are more likely to have multiple hypoplasias (17.6%) as compared to individuals with no non-utilitarian offerings (8.7%), their relationship does not explain the association between hypoplasias and age-at-death (Goodman *et al.*, 1983). However, the inability of grave goods to explain the observed association is probably more a function of their uncertainty as indicators of status than of the insignificance of status differences in the aetiology of childhood stress and adult mortality.

Hammon–Todd: LEH, cause of death and longevity

The relationship between LEH and adult morbidity has also been studied in the Hammon–Todd Osteological Collection. The advantages of this historical collection as compared to archaeologically derived collections include the availability of data on country of origin, documented age at death, and documented cause of death (from death certificates).

The Hammon–Todd collection, housed at the Cleveland Museum of Natural History, is comprised of individuals who died during the earlier part of this century. They are from the lower socioeconomic segment of the greater Cleveland population. The sample selected is limited to individuals over 18 years of age with birth certificates or other documents which reliably establish birth dates and, therefore, age at death. Additional selection was made for individuals with at least four different anterior teeth with minimal attrition. The study sample consists of 185 individuals who were born between 1852 and 1912 and died between 1915 and 1931. The mean age at death is 37.7 years (range 18 to 67 years).

Defects were recorded on the anterior teeth, either on the left or right side, depending on which was best preserved. As in the Dickson Mounds study, presence or absence of the tooth was scored, estimation was made of enamel lost to study through attrition or other destructive process, and the

Table 11.2. *Comparison of mean ages at death of individuals with and without linear enamel hypoplasias on mandibular anterior teeth*

Tooth	No LEH Mean (S.D.)	LEH Mean (S.D.)	F Ratio	P 2-tail
I₁ (1–4 yrs)*	38.1 (11.4) n=84	33.4 (11.3) n=43	4.42	.037
I₂ (1–4 yrs)*	38.4 (11.7) n=107	34.3 (11.8) n=39	3.35	.069
C (2–6.5 yrs)*	37.8 (11.1) n=74	35.0 (11.9) n=81	2.74	.098

*Years in parenthesis refer to the approximate ages covered by the enamel portion studied.

location of defect was used to estimate the individuals' age at formation of the defect (Goodman & Armelagos, 1988).

Individuals with LEH have a significant decrease in life expectancy (Table 11.2). For example, for the mandibular anterior teeth, there is between a 2.8 (canine) and a 4.7 (central incisor) year decrease in life expectancy with one or more LEHs (Note that the estimated beginning period of development of each tooth has been excluded due to attrition and that the teeth span different developmental periods.)

Table 11.3 provides an analysis of the relationship between mean ages at death of individuals with and without LEH by three annual periods on the mandibular central incisor. Individuals with LEH between 1 and 2 years live slightly, though insignificantly, longer. However, in comparing individuals with and without LEHs between 2 and 3 years, and 3 and 4 years, one finds that individuals with LEH die at a mean earlier age. Individuals with LEH between 2 and 3 die at a mean age of 33.4 years, while those without LEH between 2 and 3 die at a mean age of 39.1 years. This 5.7 year difference is significant at the 5% probability level. Similarly, individuals with LEH between 3 and 4 die at a mean age of 29.6 years, while individuals without LEH between 3 and 4 die at a mean age of 38.3 years (Table 11.3).

The relationship between age at death and age at formation of a LEH is further shown in this comparison of the frequency of LEH by time of occurrence for individuals who died before 30 years of age and those who died after age 30. Differences in the frequency of LEH between these two groups are insignificant up to two years of age. However, between 2.0 and 2.5 years, nearly three times as many of the under-30 group have LEHs as compared to the over-30 group. This increased incidence of defects continues to at least four years of age or the end of the tooth crown's

Table 11.3. *Comparison of mean ages at death of individuals with and without linear enamel hypoplasias by yearly developmental zones on the mandibular central incisor*

I_1	No LEH Mean (S.D.)	LEH Mean (S.D.)	F Ratio	P 2-tail
1–2 years	36.2 (11.4) $n=114$	38.5 (14.6) $n=13$	0.47	ns
2–3 years	39.1 (12.0) $n=113$	33.4 (11.4) $n=33$	5.90	.016
3–4 years	38.3 (11.8) $n=142$	29.6 (12.4) $n=8$	4.01	.047

development. Even though the first year is a period of rapid brain size increase, these data suggest that time periods after the first year can have consequence for longevity (Figure 11.4).

Lastly, mean ages at death are compared for individuals with and without LEH and divided into four common causes of death: infectious disease (usually tuberculosis), heart disease, cancer, and alcoholism (Table 11.4). Sample sizes are quite small. None the less, it is interesting that the greatest effects are seen in the infectious disease and alcoholism groups. Individuals who died of an infectious disease and had a hypoplastic defect died at a mean of 31.2 years, compared to 35.1 years for those who died of an infectious disease without a LEH (differences = 3.9 years). Individuals who died of alcoholism and with a LEH died at a mean age of 31.5 years, or 7.9 years before those who died of alcoholism but without a LEH (Table 11.4). These data suggest that decreased survival with adult infections is perhaps related to weakened immunity or bouts of infectious disease in early life, as evidenced by LEH.

Implications and conclusions

The data presented suggest that metabolic challenges (stresses, perturbations) around ages two to seven that are severe enough to cause a disruption to amelogenesis (and a LEH) are consistently associated with earlier ages at death. This 'postweaning' period is a relatively unexplored time period, and one in which consistent health records are hard to locate consistently.

The main question deriving from the archaeological studies concerns the biocultural processes by which these diverse phenomena are associated. The Dickson data most strongly supports the cultural buffering hypothesis. However, this mechanism is not exclusive, and the data on the association between LEH and idiopathic brain damage suggest that some form of wear

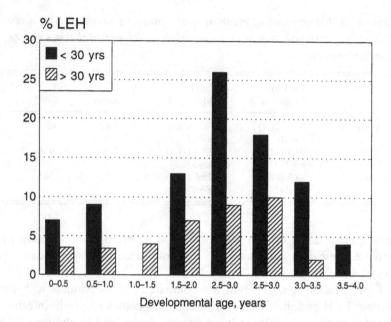

Figure 11.4. Comparison of the percentage of linear enamel hypoplasias (LEH) by half-year developmental periods for individuals who died before and after their thirtieth birthday.

and tear or biological damage mechanism might also be implicated. This is also consistent with the infectious disease results from the Hammon–Todd study.

LEH appears to be an effective tool for studying the longitudinal pattern of stress, morbidity, and mortality in past and living populations. It seems to be sensitive to developmental disruptions, is associated with a number of types of stress, is indelible and its age at formation can be reliably estimated.

Enamel hypoplasias and other biological markers may provide an alternative means for studying the long-term effects of stress in contemporary populations. Since these markers provide a biologically unbiased 'memory' of stress, they may yield more consistent patterns of effect than individuals' recall of stressful events or other retrospective measures of exposure. Whereas this paper has focused mainly on methodological questions, the data beg further elaboration of theory and models that might help to better contextualize data linking early developmental events and adult health.

Table 11.4. *Comparison of mean ages at death of individuals with and without linear enamel hypoplasias on the mandibular central incisor divided by cause of death (COD)*

I$_1$	No LEH Mean (S.D.)	LEH Mean (S.D.)	F Ratio	P 2-tail
Infectious	35.1 (10.7) $n = 48$	31.2 (9.5) $n = 30$	2.65	ns
Coronary	43.9 (14.6) $n = 16$	41.3 (18.3) $n = 6$	0.12	ns
Accident	39.7 (10.4) $n = 13$	37.8 (10.2) $n = 5$	0.13	ns
Alcoholism	40.4 (7.4) $n = 7$	31.5 (4.9) $n = 2$	2.45	ns

References

Anderson, B.G. & Stevenson, P.H. (1930). The occurrence of mottled enamel among the Chinese. *Journal of Dental Research*, **10**, 233–8.

Barker, D.J.P. & Osmond, C. (1986a). Childhood respiratory infection and adult chronic bronchitis in England and Wales. *British Medical Journal*, **293**, 1271–5.

Barker, D.J.P. & Osmond, C. (1986b). Infant mortality, childhood nutrition, and ischaemic heart disease in England and Wales. *Lancet*, **i**, 1077–81.

Barker, D.J.P. & Osmond, C. (1987). Death rates from stroke in England and Wales predicted from past maternal mortality. *British Medical Journal*, **295**, 83–6.

Barker, D.J.P., Bull, A.R., Osmond, C. *et al.* (1990). Fetal and placental size and risk of hypertension in adult life. *British Medical Journal*, **301**, 259–62.

Barker, D.J.P., Godfrey, K.M., Osmond, C. & Bull, A. (1992). The relation of fetal length, ponderal index and head circumference to blood pressure and the risk of hypertension in adult life. *Pediatric and Perinatal Epidemiology*, **6**, 35–44.

Chandra, R.K. (1975). Fetal malnutrition and postnatal immunocompetence. *American Journal of Diseases of Childhood*, **129**, 450–4.

Clark, G.A., Hall, N.R. & Armelagos, G.J. *et al.* (1986). Poor growth prior to early childhood: decreased health and life-span in the adult. *American Journal of Physical Anthropology*, **70**, 145–60.

Cohen, J. (1974). Population differences in dental morphology viewed in terms of high and low heritability. *American Journal of Physical Anthropology*, **41**, 473.

Cook, D.C. & Buikstra, J.E. (1979). Health and differential survival in prehistoric populations: prenatal dental defects. *American Journal of Physical Anthropology*, **51**, 649–64.

D'Avanzo, B., La Vecchia, C. & Negri, E. (1994). Height and the risk of acute myocardial infarction in Italian women. *Social Science and Medicine*, **38**, 193–6.

Duray, S. (1994). Enamel defects and reduced age at death in prehistoric Native Americans. *American Journal of Physical Anthropology*. Suppl. 18, 83–4.

Golden, M.H.N. (1994). Is complete catch-up growth possible for stunted children? *European Journal of Clinical Nutrition*, **48** (Suppl. 1), S58–S71.

Goodman, A.H. & Armelagos, G.J. (1988). Childhood stress and decreased longevity in a prehistoric population. *American Anthropologist*, **90**, 936–44.

Goodman, A.H. & Armelagos, G.J. (1989). Infant and childhood morbidity and mortality risks in archaeological populations. *World Archaeology*, **21**, 227–42.

Goodman, A.H., Armelagos, G.J. & Rose, J.C. (1980). Enamel hypoplasias as indicators of stress in three prehistoric populations from Illinois. *Human Biology*, **52**, 515–28.

Goodman, A.H., Armelagos, G.J. & Rose, J.C. (1984a). The chronological distribution of enamel hypoplasias from prehistoric Dickson Mounds populations. *American Journal of Physical Anthropology*, **65**, 259–66.

Goodman, A.H., Lallo, J., Armelagos, G.J. & Rose, J.C. (1984b). Health changes at Dickson Mounds, Illinois (AD 950–1300). In *Paleopathology at the Origins of Agriculture*. ed. M.N. Cohen and G.J. Armelagos, pp. 271–305. New York: Academic Press.

Goodman, A.H., Martinez, C. & Chavez, A. (1991). Nutritional supplementation and the development of linear enamel hypoplasia in children from Tezonteopan, Mexico. *American Journal of Clinical Nutrition*, **53**, 773–81.

Goodman, A.H., Pelto, G.H., Allen, L.H. & Chavez, A. (1992). Socioeconomic and anthropometric correlates of linear enamel hypoplasia in children from Solis, Mexico. *Journal of Paleopathology*, Monograph 2, 373–80.

Goodman, A.H. & Rose, J.C. (1990). Assessment of systemic physiological perturbations from dental enamel hypoplasias and associated histological structures. *Yearbook of Physical Anthropology*, **33**, 59–110.

Goodman, A.H. & Rose, J.C. (1991). Dental enamel hypoplasias as indicators of nutritional status. In *Advances in Dental Anthropology*, ed. M.A. Kelly and C. Larsen, pp. 279–93. New York: Wiley-Liss.

Goodman, A.H., Rothschild, N.A. & Armelagos, G.J. (1983). Social status and health in three prehistoric populations from Dickson Mounds, Illinois. *American Journal of Physical Anthropology*, **60**, 199.

Harn, A. (1971). *The Prehistory of Dickson Mounds: A Preliminary Report*. Springfield Illinois: Illinois State Museum.

Harn, A. (1978). Mississippian settlement patterns in the central Illinois River Valley. In *Mississippian Settlement Patterns*, ed. B. Smith, pp. 233–68. New York: Academic Press.

Harn, A. (1980). *The Prehistory of Dickson Mounds: The Dickson Excavation*. Springfield Illinois: Illinois State Museum, report No. 36.

Harn, A. (1986). The Eveland Site: Inroad to Spoon River Mississippian Society. Paper presented at the Annual Meetings of the Society for American Archaeology, April 23–27, New Orleans.

Herrenkohl, L.R. (1979). Prenatal stress reduces fertility and fecundity in female offspring. *Science*, **206**, 1097–9.

Jaffe, M., Attias, D., Dar, H., Eli, I. & Judes, H. (1985). Prevalence of gestational and perinatal insults in brain-damaged children. *Israeli Journal of Medical Science*, **21**, 940–4.

Kreshover, S. (1960). Metabolic disturbances in tooth formation. *Annals of the New York Academy of Science*, **85**, 161–7.

Lallo, J., Armelagos, G.J. & Mensforth, R.P. (1977). The role of diet, disease and

physiology in the origin of porotic hyperostosis. *Human Biology*, **49**, 471–83.
Lallo, J., Armelagos, G.J. & Rose, J.C. (1978). Paleopathology of infectious disease in the Dickson Mounds Population. *Medical College of Virginia Quarterly*, **14**, 12–23.
Magennis, A. (1990). *Growth and transverse line formation in contemporary children.* Unpublished PhD Dissertation, Amherst, Massachusetts: University of Massachusetts.
Martin, D.L., Goodman, A.H. & Armelagos, G.J. (1985). Skeletal pathologies as indicators of quality and quantity of diet. In *The Analysis of Prehistoric Diets*, ed. R.I. Gilbert and J. Mielke, pp. 227–9. New York: Academic Press.
Marmot, M.G., Shipley, M.J. & Rose, G. (1984). Inequalities in death – specific examinations of a general pattern? *Lancet*, 1003–6.
Martorell, R., Rivera, J. & Kaplowitz, H. (1990). Consequences of stunting in early childhood for adult body size in rural Guatemala. *Annals Nestlé*, **48**, 85–92.
Massler, M., Schour, I. & Sarnat, B. (1941). Developmental pattern of the child as reflected in the calcification pattern of the teeth. *American Journal of Diseases of Childhood*, **62**, 33–67.
Miler, I. (1982). Nutrition in early life and the development of resistance and immunity. *Biblitheca Nutritio et Dieta*, **31**, 55–60.
Mittler, D.M. & Van Gerven, D.P. (1994). Developmental, diachronic, and demographic analysis of cribra orbitalia in the Medieval Christian populations of Kulubnarti. *American Journal of Physical Anthropology*, **93**, 287–97.
Moore, J., Swedlund, A.C. & Armelagos, G.J. (1975). The use of life tables in paleodemography. *American Antiquity*, Memoir No. 30, 57–70.
Mosley, W.H. (1983). Child survival: research and policy. *Population and Development Review*, Supplement to volume 10, 3–23.
Nie, N.H., Hull, C.H., Jenkins, J.G., Steinbrenner, K. & Bent, D.H. (1975). *SPSS Statistical Package for the Social Sciences.* 2nd edn. New York: McGraw Hill.
Pindborg, J.J. (1970). *Pathology of the Dental Hard Tissues.* Philadelphia: Saunders.
Pindborg, J.J. (1982). Aetiology of developmental enamel defects not related to fluorosis. *International Dental Journal*, **32**, 123–34.
Preston, S.H. (1980). *Biological and Social Aspects of Mortality and Length of Life.* Liege: Ordina.
Rose, J.C., Armelagos, G.J. & Lallo, J.W. (1978). Histological enamel indicators of childhood stress in prehistoric skeletal samples. *American Journal of Physical Anthropology*, **49**, 511–16.
Rothschild, N.A. (1979). Mortuary behavior and social organization at Indian Knoll and Dickson Mounds. *American Antiquity*, **44**, 658–75.
Sarnat, B.G. & Schour, I. (1941). Enamel hypoplasias (chronic enamel aplasia) in relationship to systemic diseases: a chronological, morphological and etiological classification. *Journal of the American Dental Association*, **28**, 1989–2000.
Schmolling, P. (1984). Human reactions to the Nazi concentration camps: a summing up. *Journal of Human Stress*, **10**, 108–20.
Stein, Z.A., Susser, M.W., Saenger, G. & Marolla, F.A. (1975). *Famine and Human Development: The Dutch Hunger Winter of 1944/45.* New York: Oxford University Press.
Via, W.F. & Churchill, J.A. (1959). Relationship of enamel hypoplasia to Abnormal Events of Gestation and Birth. *Journal of the American Dental*

Association, **59**, 702–7.

Wadsworth, M.E.J. (1986). Serious illness in childhood and its association with later-life achievement. In *Class and Health: Research and Longitudinal Data*, ed. R.G. Wilkinson, pp. 50–74. London: Tavistock.

Waaler, H.T. (1984). Height, weight and mortality. The Norwegian experience. *Acta Medica Scandinavia* (Suppl.), **679**, 1–56.

White, T. (1978). Early hominid enamel hypoplasia. *American Journal of Physical Anthropology*, **49**, 79–83.

Wood, C.S. (1983). Early childhood, the critical stage in human interactions with disease and culture. *Social Science and Medicine*, **17**, 79–85.

Yaeger, J. (1980). Enamel. In *Orban's Oral Histology and Embryology*, 9th edn. ed. S. Behaskars, pp. 46–106. St. Louis: CV Mosby.

12 *The childhood environment and the development of sexuality*

M.P.M. RICHARDS

Introduction

There is an old tradition in biosocial research on sexual behaviour of exploring correlations between aspects of physical development (most often puberty) and aspects of social and sexual development. When relationships are found, they are usually presented as evidence for a common-sense view that the timing of puberty determines the beginning of sexual interest and so the start of sexual behaviour (e.g. Hurlock, 1995). These, in turn, it is suggested, lead to the establishment of long-term relationships and marriage. However, many social scientists reject such biological determinist views of social and sexual development and relationship formation. Their approach to love, sex and marriage has usually been very much more social problem orientated, at least in the quantitative survey traditions, and has been dominated by studies of teenage pregnancy and contraception (e.g. Miller & Moore, 1990). The chief concern has been to see who begins to do what, when, and with what consequences (principally conception) (see Griffin, 1993). Between these two approaches, there is very little of what might be termed a developmental understanding of how young people enter the world of adult sexual activities and relationships. One might expect this gap to be filled by developmental psychology. But examination of textbooks of this field reveal a curious silence. There are almost no accounts of the development of sexuality. Instead, we find discussions of gender identity and often an account of the physical and physiological changes that occur at puberty. The implicit assumption of this approach seems to be that having established a male or female gender identity, sexual behaviour will emerge as an inevitable result of the biological changes seen at puberty. So here, once again, we have another version of biological determinist model which has been the biosocial tradition.

The absence of any elaborated theory of sexual development encourages

183

184   *M.P.M. Richards*

the perpetuation of biological determinist models and of the social problem orientated approach of the social scientists. But it could be argued that neither the biological determinist position nor the head counting approach increase our understanding of the development of sexuality. What does the percentage of 16 year-olds who have experienced sexual intercourse tell us about sexual development? Would we think differently about sexual development if the figure was 20% rather than 40%? Such figures may influence social policy related to such matters as sex education or contraceptive availability but they tell us little about sexuality (see Lewontin, 1995 for a biologist's critique for such approaches). Most obviously very little is learnt about the meaning to young people of the activities they have taken part in, or avoided. Much, but not all, sexual activity takes place within the context of relationships. How does the quality of these and their power dynamics relate to sexual activity? The stark facts of the reported age of first intercourse, its frequency and, perhaps the number of partners over a period of time, give us few clues to answer such questions.

   The aim of this chapter is to begin to put together at least the elements required for a theoretical understanding of the development of sexuality. Given the nature of the field, current understanding is fragmentary. However, there are encouraging developments in social research which are beginning to provide some rather more satisfactory exploration of this important territory. Building on this new research, a theoretical view of sexuality will begin to be constructed which sees its development as a long-term process with its origins early in development and continuing on through the life course into adulthood. But before discussing these recent developments and building on them to begin to construct a developmental approach, some aspects of the biosocial and survey approaches will be examined in a little more detail.

A biological story

It is well established that the childhood experience of parental divorce or separation is later associated with earlier marriage (or cohabitation) and parenthood. The observation appears to be a robust one and has been found in the large British cohort studies (e.g. Maclean & Wadsworth, 1988; Kuh & Maclean, 1990; Elliott & Richards, 1991; Kiernan, 1992) as well as in North American work (e.g. Amato, 1993; Amato & Keith, 1991). How might this association be explained? One approach would be to draw on research that shows that children who have experienced parental divorce may grow up socially a little faster (Weiss, 1975). The experience of living in a single parent household may weaken generational boundaries and lead to an earlier social maturity and more adult attitudes at earlier ages. For some young people, a parental remarriage and conflicts with a step-parent may lead to a desire to leave home at an earlier age and this may encourage the setting up of cohabitations or marriage. Another factor that could be involved is a tendency to leave school earlier after a parental divorce, which once again may encourage an earlier transition to adulthood.

Others have looked in rather different directions for their explanations of the link between parental divorce and earlier marriage. It has been found in both North American and European studies that parental divorce is associated with a younger average age at first heterosexual intercourse (Simmons & Blyth, 1987; Stattin & Magnusson, 1990; Newcome & Udry, 1987; Small & Lusher, 1994). Age at first intercourse is associated with age at marriage and both of these have been associated with age at puberty especially for women (Udry & Cliquet, 1982; Sandler, Wilcox & Horney, 1984; Kiernan, 1977; Miller & Heaton, 1991; Udry & Billy, 1987).

Another link to this chain is provided by a New Zealand study which found earlier menarche for girls who had experienced father absence in childhood (Moffitt *et al.*, 1992). These associations were investigated in the context of a sociobiological hypothesis produced by Belsky and others (1991). Their claim was that social conditions in childhood would (at least for girls) determine one of two possible marriage and child-bearing strategies. If conditions in childhood are relatively conflict free and stable and the parental marriage persists, the supposition is that young women will delay their own marriage and child-bearing. They will follow the (typically middle-class) pattern of prolonging education and delaying partner choice and marriage. In sociobiological terms, this is a K strategy in which small numbers of children are produced, but in which each is accorded a high rate of parental investment. When childhood conditions are less stable, with more conflict, Belsky *et al.*, suggest that young women follow something more like an R strategy: that is to say they begin child bearing early and produce many children, not necessarily in the context of a continuing relationship with a father. This strategy of low parental investment is, by analogy, what is supposed to characterize the underclass and many black populations in North America. According to Belsky *et al.*, the switch which determines the strategy that young women will adopt is the age at menarche. In effect, their claim is that the strategy that young women will adopt, either early and frequent child bearing with low parental investment or later widely spaced births with high parental investment, will be determined by the age of menarche, early in the first case and later in the second.

There has been one test of the Belsky *et al.* (1991) hypothesis which uses data from a New Zealand longitudinal study (Moffitt *et al.*, 1992). This found no evidence at all for the proposition that age at menarche was related to family stress or behavioural problems in childhood. However, an association was found between father absence in childhood and earlier menarche. This latter finding is in line with an earlier retrospective survey from the USA (Surbey, 1990) and a study of adolescence which links earlier puberty to reported parent–child social distance (Steinberg, 1988).

Moffitt et al. (1992) in rejecting the Belsky et al. (1991) sociobiological model, propose instead a genetic transmission model. For this, they suggest that age at menarche is heritable (there is some evidence of a correlation between timing for mothers and daughters, Garn, 1980) and then, drawing on the evidence cited earlier in this chapter, they suggest a chain running from earlier menarche to earlier sexual intercourse, marriage and earlier births and earlier divorce. Thus a link between parental divorce and a child's age at marriage, and so divorce, is made, and it is suggested that this is indirectly mediated by age at menarche, which these authors assume to be heritable. This explanation adopts the traditional biosocial position that holds that the age at puberty determines subsequent social and cultural behaviour. But the authors do state that their explanation 'should not discourage further enquiries about the psychosocial context of pubertal development' (Moffitt et al., 1992, p. 56). While it is clear that the Belsky model should be rejected (and with it the inherent sociobiological and political assumptions) as it has no empirical support, the model does have the virtue of suggesting, as others have done previously, that puberty should not be viewed as a biologically determined event. Rather, it may be seen as a developmental stage whose timing may depend on earlier social relationships, among other factors. Contrary to general belief, the factors associated with the timing of puberty 'are complex and poorly understood' according to a recent review (Hopwood et al., 1990). This point will be returned to in discussion of the development of sexuality.

Using British data (Richards & Elliott, 1993) the possible link between parental divorce and age at puberty has been examined. With data from the National Child Development Survey (a National cohort of some 17 000 children born in 1958) no association was found between earlier parental divorce and age of puberty for women or men. This finding casts doubt on the generality of the Moffitt et al. (1992) explanation. If timing of puberty is heritable in a New Zealand population (largely derived from the UK by emigration) the same might be expected to be true in the British sample given the cultural and genetic similarities between the populations. Part of the other link in the Moffitt et al. hypothetical chain can be demonstrated in the UK. British data (e.g. Kiernan, 1977) suggests that earlier puberty is associated with earlier marriage or cohabitation but not with earlier child-bearing for women. For women, but not for men, the association was

Table 12.1. *Percentage cohabiting or married or having a child before the age of 20 and age at puberty*

Age of puberty	Cohabiting or marriage <20		Childbearing <20	
	Women	Men	Women	Men
11 years of under	32.7%	11.5%	14.7%	3.1%
15 years or older	24.9%	6.8%	16.8%	3.1%
	p<0.001	p<0.001	p<0.001	ns

From Richards and Elliott, 1993.

also found with parenthood before the age of 20, but this indicated that those having a baby before this age were overrepresented in the late menarche group. (see Table 12.1)

How are these associations between age of puberty and age at marriage and child-bearing to be understood? The usual interpretation, as already mentioned, is to assume a direct causal pathway involving hormonal changes following puberty. But, there are many other ways in which the association could arise. There could be a third factor, perhaps particularly social attitudes and behaviour, which might independently influence both the age at puberty and age at marriage. Other possibilities might involve changing notions of self-identity arising from the perception of the bodily changes at puberty, the attitudes of others toward these and social actions which may follow from these. There is a considerable body of psychological research which associates earlier and later puberty with a number of psychological correlates (e.g. Ruble & Brooks-Gunn, 1982; Greif & Ulman, 1982; Simmons & Blyth, 1987; Stattin & Magnusson, 1990).

Turning to evidence that might bear on the postulated direct link between puberty and sexual behaviour, a study by Gagnon (1983) of about 500 women in American colleges found 'a rather limited relationship between menarche and youthful sexuality'. Those women with the earliest menarche were slightly more likely to date and pet in High School and subsequently began intercourse slightly earlier than those with later menarches. Gagnon concludes that 'a strong case cannot be made for a biological argument ... particularly since the masturbation data show no differences' [between those with early and late menarche] but he does acknowledge some methodological weaknesses in his study, especially the retrospective nature of the data. As with other studies, a correlation between age at puberty and the initiation of heterosexual behaviour is apparent, but relatively weak, (e.g. Stattin & Magnusson, 1990), and any such correlation is open to a wide range of interpretations other than a simple causal hormonal hypothesis.

In a series of studies of young people in the USA, Udry and his colleagues have attempted to explore directly possible links between the hormonal changes at adolescence and sexual behaviour. In an early paper (Udry, Talbert & Morris, 1986), they set out to distinguish a biological hypothesis, that hormonal changes associated with puberty increase libido and so stimulate sexual behaviour, and a 'sociological' model which postulates that the hormonal changes lead to pubertal development which is then 'socially interpreted as a signal for social encouragement of sexual behaviour'. Using questionnaires which asked about both sexual behaviour and feelings and assays of hormones in serum samples, groups of young men and women were studied. Udry and his co-workers found some significant hormonal level (primarily androgens) correlations with some aspects of sexual behaviour and sexual feelings, but not with the beginning of sexual intercourse. They suggested that these correlations with sexual behaviour for boys and girls are 'direct', i.e. that they do not operate via pubertal development or as intermediate variables. The exception to this, they suggest, is masturbation in girls which does relate to these latter two variables. Unlike in girls, occurrence of intercourse in boys was related directly to androgen levels. From these results, the authors conclude that there is support for a mixed biological and sociological model (see also Udry, 1988).

Subsequently, Udry and Billy (1987) reported results from a larger panel study of the population of whole schools. Data were collected from over a thousand young people at mean ages of 13.6 and 15.4 years. The primary interest in the study was the transition to beginning of sexual intercourse between the two age-points in relation to reported motivation, social controls, attractiveness and hormonal changes. For white males, beginning of intercourse was associated with hormonal levels and social attractiveness, while for the white girls transition was most closely correlated with the social control measures, with no link to the attractiveness measure, hormone level or reported sexual motivation. For black females, pubertal development seemed the key variable. Data for black males were not reported because most had already had sexual intercourse at the start of the study, and so in many cases before they had reached puberty (see also Udry, 1990).

The results of these studies do not support any simple causal biological hypothesis. They demonstrate that there are associations between aspects of sexuality, hormonal levels, pubertal development, age and social context, but that the links between these are complex. There are clear differences between boys and girls in the pattern of association and between the black and white groups. The data indicate that puberty in terms of bodily change or the hormonal changes is neither necessary or sufficient for the development of sexual intercourse. Just as we can rule out any simple causal biological hypothesis, however, we can discount any framework which deals solely in terms of social or psychological factors. Our notions of the development of sexuality need to bring together both these realms and they seem likely to be culturally specific, as the Udry *et al.* data strongly suggests, we will need rather different explanations for boys and girls in the black and white communities (in the USA, see also Furstenberg *et al.*, 1987) and, presumably, similar variations will occur elsewhere. As will be seen below in the discussion of the surveys of sexual behaviour, we also need accounts which allow for the considerable historical changes in recent decades in the patterns of sexual behaviour and in the ages at which these develop. As a final comment, it is important to emphasize that, whatever may be the links between puberty and the start of heterosexual intercourse, the evidence we have cannot on its own provide any explanations for the positive correlations between age at menarche and age at marriage, and the negative one between menarche and the start of child-bearing.

Surveys of sexual behaviour and other social research

From the 1930s onwards, a large number of surveys of sexual behaviour have been carried out in North America, Britain and other parts of the world. The motives for carrying out such research have varied. The most famous of all such surveys, that of Kinsey (Kinsey, Pomeroy & Martin, 1948; Kinsey *et al.*, 1953), was carried out to provide public information about private behaviour to inform sex education and the population more generally. Often, however, research has been driven by concerns about social problems especially 'premature' sexuality and teenage pregnancy. More recently, the spread of the HIV virus has led to a new generation of surveys. Trying to establish what people actually do, through asking them directly, has been a much stronger motivation than concern about theoretical issues related to sexuality in this tradition of research. While Kinsey adopted a simple hydraulic model of sexual behaviour which counted 'outlets' and their frequency, the more social problem-orientated surveys, not surprisingly, concentrated on occurrence of the aspects of behaviour seen to be most problematic. Thus Schofield's (1968, 1973) British survey of the sexual behaviour of young people concentrated on sexual intercourse and the use of contraception as these were the issues of concern. He did not attempt to investigate wider aspects of sexuality, for example, sex with same sex partners or solitary sexual activities. There was little interest in the social context in which sexual intercourse might occur, or the social meaning of the behaviour to the participants. Recently, though it is still unusual, there have been a few surveys which are designed within a theoretical account of sexuality (e.g. Laumann *et al.*, 1994). This body of survey research has been reviewed frequently (e.g. Wight, 1990; Moore & Rosenthal, 1993; Meyrick & Harris, 1994; Miller, Christophersen & King, 1993; Johnson *et al.*, 1994).

With all this survey research, as with all the studies of sexual behaviour that have been mentioned in this chapter, there are important issues of reliability and validity. How accurately do the respondents to surveys and other research report what they have done? There is internal evidence in many surveys that the gap between what people tell interviewers and what they do may be significant. Men tend to report more sex with more partners than women. Individuals, such as prostitutes, who may have very different patterns of sexual activity than the majority of the population very seldom turn up amongst survey respondents. Response rates to surveys tend to be very low with anything over 60% being considered as very good, raising questions about their representativeness of the samples for all this research. If we take this work at its face value, we see a changing pattern in Britain with a declining age at first sexual intercourse and a growing minority of both boys and girls having intercourse before the legal minimum of 16. There are associations with social class and education with middle-class young people and those remaining in education having later first experiences of intercourse (e.g. Ford, 1991; Johnson *et al.*, 1994; Knox, MacArthur & Simons, 1993). The few data that exist suggest differences in patterns between ethnic communities, as well as geographical variation. Scottish data, for instance, suggest a somewhat later start of intercourse north of the border (West, Wight & Macintyre, 1996). Because of concerns about the HIV virus, as well as growing awareness of homosexuality, it is increasingly common for surveys to enquire about same-sex partners. Broadly speaking, there is a historical trend with younger generations reporting a wider range of sexual activities (though less use of prostitutes). For any one generation, those of higher social class or higher levels of education report a greater range of activities, but it is important to emphasize that these differences may have as much to do with what people in various groups are prepared to discuss with interviewers as with variation in their sexual activities.

Provided there is reasonable confidence that the broad trends in behaviour that surveys describe do represent variation in behaviour rather than in reporting, the surveys may be taken to describe variation in sexual activity that require explanation. However, they seem to have contributed very little to understanding of the development of sexuality and sexual behaviour. This is partly because they have been concerned with a relatively narrow range of sexual behaviour rather than sexuality more broadly defined. With a few honourable exceptions, they are not concerned with individuals as social actors so they have little to contribute about the experience of sex, sexual desire or the social context in which these take place. However, there is a small but growing body of research which has begun to explore the subjective experience of sex and sexual desire through qualitative research techniques. Examples of such research are the Women Risk and Aids Project in Britain (e.g. Holland *et al.*, 1991; Thomson & Scott, 1991) and that of Tolman, 1991, 1994 and Thompson (1990) and others in the USA. Drawing on these and wider theoretical frameworks, a development view of sexuality will now be outlined.

Theorizing the development of sexuality

In this section a number of issues will first be outlined which arguably are essential elements in a theory of the development of sexuality and then, in conclusion, there will be a brief discussion building on these elements of female and male development through adolescence. It may be noticed that discussion is not organized around an opposition of social and biological factors. An underlying theme for the argument is that this conceptualization is false and has served to mislead and distort developmental perspectives (see Gray, 1992, for example). The five issues to be discussed are (i) developmental origins; (ii) social contexts: (iii) gender identity; (iv) bodies and (v) sexual scripts.

Developmental origins

There is a common, but I believe misleading, tendency to regard puberty as the starting point for sexual development: 'Puberty's onset marks a beginning of adult sexual development' as one recent text puts it. Such approaches not only ignore earlier development, but by the prominence they give to puberty (as a biological event) may encourage biological determinist accounts. The development process can be taken to begin much earlier: at birth, conception, with a thought in the mind of either or both parents, or whenever it is chosen to begin an account of an individual's development. At least from the time of Freud, some have regarded children, or indeed infants, as sexual. The roots of adult sexuality may indeed be found in childhood but to regard all childhood activities that involve genitals as sexual, in the same way as they would be if adults did similar things, is to adopt a very adult perspective. It also misses the point that adult reactions to the ways in which children touch and expose parts of their bodies is one of the ways in which children learn what is sexual (see

Jackson, 1982). In middle childhood, when children play doctors and nurses and undress and examine each other, this is sometimes described as the beginning of sexual experimentation, but to see this behaviour as sexual may be misleading. It is perhaps more accurate to regard this as an activity which satisfies an understandable curiosity about how bodies are constructed and work. Through adult reactions to such games, children perceive that there are 'secret' activities of which adults may disapprove, at least when such games are played in more public contexts. Children learn that special rules apply to their genitals, that they should not be talked about in public, that they should be hidden in most situations and that similar conventions apply to excretion. This learning may build an association that dominates much of what some children come to understand about their genitals, that excretion and the genitals are closely linked and that both may be the subject of guilt and shame. In our culture, these rules may be different for boys and girls. In studies of infants and young children, it has been found that mothers may be much more punitive in their treatment of girls who touch their genitals than they are of boys. Middle-class mothers are likely to be much more tolerant of such activities than working-class mothers (Newson & Newson, 1963). It is tempting to relate these to differences in adult sexuality.

Long before they reach puberty, boys and girls have a wide knowledge of conventions in their families and the wider social world that govern attitudes towards their bodies. Their own attitudes and feelings about their bodies are shaped by these. They have also come to understand something of intergenerational boundaries, that there are some things better not done or discussed in front of grown-ups, but may be able to be shared with their peers. For a minority who experience inappropriate or abusive behaviour from adults, these boundaries are breached. Evidence suggests that forcing boys or girls to take part in overtly sexual adult activities may have long-lasting effects on the development of their sexuality (Kendall-Tackett, Williams & Finkelhor, 1993; Laumann et al., 1994).

Puberty should not be seen, as it often is, as an event which somehow intrudes into childhood and forces the beginning of adult sexuality. Rather, it is the process of bodily development, the timing of which is influenced by social context of childhood and a child's experiences (Brooks-Gunn & Reiter, 1990), and this in turn is experienced in different ways depending on the earlier attitudes and experiences of children and their particular social world. This physical development takes place in a phase of rapid change in social attitudes and relationships (Hill, 1993).

Social contexts

Sexual behaviour is an aspect of human activity that is overwhelmingly social (Gagnon & Simon, 1973). While there may be broad similarities in sexuality and sexual behaviour between all individuals, at least within one gender, at a more detailed level there is wide variation between cultures, over historical time and between individuals in the same culture at the same point in history. As may be seen from the research on sexual activity in young people and hormonal levels, variation in hormonal level or in physical development explains very little of the variation in frequency of sexual activity or in the nature of those activities. However, while we may point to social worlds as sites where sexuality develops, there is little understanding of the processes involved. Some specific examples of processes can be given, as has already been done in the chapter, but we are very far from having anything like a full account of what is involved. The roles of friends and siblings, however, can be pointed to, for example, as influences for an individual (Rodgers & Rowe, 1990; Rodgers, Rowe & Harris, 1992). Part of the difficulty in providing a full account is that, because of the easy assumption of a biologically based process, very little effort has gone into analysing the social processes involved.

Gender identity

The divide of gender is established in the earliest years of childhood. Children experience the world as girls or as boys and are viewed and treated by others according to their gender (Maccoby, 1988; Thorne, 1993). In this sense, they grow up in different worlds. Thus we should not speak of the development of sexuality in general, rather of the development of sexuality in boys, or its development in girls.

Boys and girls have different bodies which are regarded and treated in different ways according to their gender. I have already given the example of how mothers may react differently to young boys and girls touching their own genitals. Studies suggest that a boy's external genitals are more likely to be named by parents and at an earlier age than girls' and, even at adolescence, boys are likely to have a much more extensive set of terms for their genitals than girls (Rosenbaum, 1979; Hirst, 1994). Boys handle their penis while urinating and, because it is more visible than the sexual organs of girls, it has been argued that boys perceive their genitals as a source of pride and pleasure while girls are more likely to develop a sense of shame, disgust and humiliation about theirs (Ussher, 1989).

There is a basic asymmetry in the social worlds of girls and boys. In a society where most children are nurtured and cared for by mothers, there is an identity of gender between the principal caretaker for girls, but not boys. This means that girls develop a gender identity in relation with their mothers while boys may define theirs in contrast to their mothers and in identity with their more distant fathers (Chodorow, 1978). This fundamental difference may be reflected in all social and sexual relationships, with girls and women seeing their own identity in relation to others, while boys and men perceive themselves as more separate and see their connections with others more in terms of status and power (Gilligan, 1982; Gilligan, Lyons & Hamner, 1989).

Bodies

A theory of the development of sexuality needs, as one of its elements, a view of the body, not simply as an object, but as a lived-in subject which is perceived by an individual as part of their being. Bodies are also perceived by others in the context of social norms, attitudes and relationships. These perceptions of others provide a social world within which an individual's attitudes and feelings about their body are formed. The ways in which from infancy onwards dressing and undressing, elimination, patterns of social holding and touching all serve to indicate to growing children how they may present their bodies and act in a social space has already been mentioned: what may be covered or uncovered, in what situations, what is named and talked about, what is unnamed and not mentioned, what may be touched, or not touched, by self or particular others, and so on. This social development indicates to children what will later be seen as genital or sexual parts of their own body are regarded in ways that are quite different to other parts of their body. Children come to see that even their own actions, such as touching themselves, are bound by social conventions and rules that adults may try to impose. Across the divide of gender, children learn that male and female bodies are, in some important respects, different and treated differently. Depending on family culture and the accidents of experience, children learn varying amounts about the bodies of the opposite sex. Towards late childhood as puberty approaches, children become more interested in and aware of their own bodies and those of others. They are acutely aware of the development of the secondary sexual characteristics in themselves and their peers (Stattin & Magnusson, 1990). While girls' experience of the physical changes of puberty have been widely studied (e.g. Rosenbaum, 1979; Prendergast, 1994; Ussher, 1989; Ruble & Brooks-Gunn, 1982) there has been much less research interest in boys (but see Gaddis & Brooks-Gunn, 1985). This latter study points out that, while most girls discuss the beginning of menstruation with their mothers and others, boys seldom talk to either of their parents or members of their peer group about ejaculation or other aspects of their pubertal development.

At puberty (in Britain and the USA) most boys say they would like to be taller, more muscular and heavier. Most teenage girls say they are dissatisfied with their bodies and would like to be thinner (e.g. Stattin & Magnusson, 1990). A majority have dieted in an attempt to reduce their weight (e.g. Moore & Rosenthal, 1993). As these authors comment 'an ideal female shape for the current generation is that of a "child–woman", in which mature breast development is superimposed on the prepubertal body' (see also Steiner-Adair, 1990). Studies show clear links between measures of self-esteem and perceptions of physical attractiveness for girls and with body fitness and strength for boys (Lerner, Oulos & Knapp, 1976). While some research has pointed to physical development and attractiveness as being very important in making and sustaining early relationships with men (Walster, Aronson & Abrahams, 1966), others have found that the beginning of dating seems to have more to do with age and parental control than physical development (Dombusch *et al.*, 1981).

In most accounts of physical development, priority and privilege are given to hormonal factors. This applies both to initial sexual differentiation in fetal development and to the changes at puberty. While there is no doubt that hormones play a role in the specific growth and differentiation at these (and other) times, it would be misleading to see them as primary or ultimate causes. Hormones influence the growth and differentiation of particular tissues but this occurs because these tissues have a specific sensitivity to certain hormones. These are complex interactive processes not simple causal events. In fetal development, hormones influence sensitive tissues to differentiate into recognizably male and female anatomical forms. The release of the hormones and the differential sensitivity of the tissues of male and female fetuses can be related back in the development (though this process is poorly understood) to the fertilization of an egg with a sperm which carried either an X or Y chromosome. From the time of fertilization, male and female bodies develop along different paths.

Puberty is a drawn-out and complex process of growth and development of secondary sexual characteristics (Marshall & Tanner, 1969, 1970) which includes an overall spurt in growth with the development of an adult body form, growth of internal and external sexual organs, body hair and other secondary sexual characteristics. Menarche, or the beginning of regular menstrual bleeding, is usually taken as the event in the process which is used to time puberty for girls, while for boys the breaking of the voice is most often recorded. Over the past century or so, there has been a secular trend toward earlier puberty (Eveleth, 1986). This is usually ascribed to improving nutrition but the factors associated with the timing of puberty are poorly understood (Eveleth & Tanner, 1976). As mentioned already, earlier social context and behaviour (including an individual's sexual behaviour or its absence) seem to be involved. But these have been studied much less systematically than the neuroendocrine processes (Reiter, 1986). Variation in timing is linked to rather different psychological factors in each gender (see Moore & Rosenthal, 1993; Brooks-Gunn & Reiter, 1990; Simmons & Blyth, 1985; Stattin & Magnusson, 1990; Simmons & Blyth, 1987 for a general discussion). Early maturing boys tend to be more confident, self-assured and more popular among peers. Those developing later are shorter at the time of onset of puberty, less confident, less popular and less attractive to girls. Some of these features may persist into adulthood.

For girls, the earliest maturers have poorer body images, are heavier and more likely to develop eating disorders. They are more likely to suffer psychological difficulties and lower self esteem and do poorly at school. By 16 years of age, they are more likely to be dating and see themselves as more popular with boys. Later maturing girls appear to do better in school work.

As far as reactions to the first menstruation are concerned, studies report a very wide range of reactions and much seems to depend on how far this has been discussed beforehand with the mother and others. Most girls, however, see it as significant and memorable (see Grief & Ulman, 1982; Brooks-Gunn & Reiter, 1990). By contrast, we have very little systematic study of how boys feel about the coming of any of the signs or stages of puberty. Given the evidence of studies which compare early and late maturing boys, one might expect reactions to be positive.

Sexual scripts

All surveys of sexual behaviour show that there is very wide variation in the frequency and nature of sexual activities that individuals engage in, or would like to engage in. In a classic study, Gagnon and Simon (1973) set out a concept of socially constructed sexual scripts which describe and structure an individual's pattern of sexual interest, behaviour and desire. They assume that scripts develop through childhood, through the kinds of processes that have been outlined in the previous sections.

A major variation of scripts for each gender is whether partners of the same or opposite sex are preferred. A considerable minority of both women and men engage in sexual activities with others of the same sex. Others may fantasize about this but not do it. A person's identification of themselves as homosexual or bisexual does not automatically follow from sexual activity with a same sex partner. Especially early in sexual development, same sex activity may be quite frequent, and it does not necessarily lead to homosexual self-identification or interest in same sex partner later on in adulthood. In cultures which are strongly heterosexual, there may be some delay before individuals see themselves as homosexual, and bisexual identity may form even more slowly (e.g Rust, 1993).

Sexual scripts not only concern preferred sexual activities, whether heterosexual or homosexual, but also sexual feelings and desires. Such feelings do not arise spontaneously but are focused and named through childhood and youth, in part through cultural expectations. As in so many things, there seems to be wide gender difference, with the process of development being more variable and remaining less clearly focused for longer in girls (Dusek, 1991). Part of this difference may relate to cultural expectations which see sexual desire developing spontaneously and much earlier and more powerfully in males, while female desire is held to be something only brought into existence through sexual activity with a male. According to the ideology, young women want relationships not sex. There may be sanctions against women, especially young women, speaking of sexual desire and it is seldom mentioned in sex education or other public discussions (Fine, 1988; Tolman, 1991, 1994). Research that has moved beyond a concern simply with what people do, to take account of what they feel about what they are doing shows, for example, strong links between mothers' attitudes to their daughters' developing bodies, their openness to discuss sexuality with them, their positive view of sexuality and their daughters own sexual experience (Thompson, 1990). Such work illustrates some of the processes by which sexual scripts related to sexual activity and desire may be written. Some have claimed that desire is partly shaped by the words available to describe it and sexual activity more generally (Hirst, 1994).

Becoming sexual

In the final section of this chapter, the process of sexual development through adolescence for girls and boys is outlined as a way of bringing together the issues that have been considered in the previous section.

Young women

For many women, adolescence is the time when the knowledge, attitudes and feelings about their own body are beginning to be deployed in heterosexual social encounters. There are age-related pressures from peer groups to engage in heterosexual relationships. The physical development of the adolescent body may attract attention from males as well as being seen by the young women as a maturity that readies them for a heterosexual relationship. Much of the focus is on relationships, intimacy and love. There are ample models for romantic relationships in western culture, in books and magazines, in films and the conversation of female friends. What girls are expected to seek is love and not sex. What she may feel may be another matter (Tolman, 1994).

Over recent years there have been several studies of teenage cultures which put great emphasis on reputation as being an overriding concern for girls (McRobbie, 1978; Griffin, 1985; Cowie & Lees, 1987). These suggest that girls must tread a narrow path with men, between being seen as a 'slag' or a 'drag'. Such social pressures enter into the negotiation of contraceptive use and safe sex within a relationship (Holland et al., 1990, 1991). As women's feelings are supposed to be spontaneous, and sexual activity something that 'just happens', a girl preparing herself by bringing a condom may indicate she is 'easy'. So the practice of safe sex may for her involve risking losing her reputation.

There are many studies that describe the progression of social and sexual activity from dating to sexual intercourse (see, for example, Chilman, 1986; Moore & Rosenthal, 1993; Miller et al., 1993; Wight, 1990). This progression is paralleled by a changing pattern of confiding relationships. In the early teenage years, these are largely confined to female friends and mothers. By the age of 19 years, it is male partners who are most likely to occupy this role (Monck, 1991). What such studies less often make clear is that, for some, an activity that becomes sexual may long predate heterosexual (or homosexual) relationships. Data suggest that, though fewer young women masturbate than young men (for whom it is more or less universal), women may start at younger ages (Moore & Rosenthal, 1993). Studies of the experience of first and early sexual intercourse suggest that there is a wide variety in ways in which this is experienced. Many, or most, do not find early experiences very pleasurable. But, those who do are much more likely to have had experience of masturbation, as well as have had sexuality represented to them by their mothers as pleasurable (Thompson, 1990). Such studies emphasize the continuity of the development of sexuality through childhood, and adolescence into adulthood. They also indicate that female experience was very varied (and apparently more so than for men). While some women may report masturbation from early childhood and experience of orgasm well before puberty, for others a first experience of orgasm may come after years of heterosexual intercourse. Masturbation, too, may not begin until well into adulthood after years of varied sex with partners. Observation of these differences, together with the few studies we have of female desire, suggest this too is very variable in its development. Some women report strong sexual feelings from childhood while, for others, it is a process of becoming more focused through sexual activity in adolescence and early adulthood. Sexual desire, love and romance are closely linked, at least in terms of cultural expectation. Experience may be rather different (Tolman, 1994). As might be predicted from the cultural expectation, most women report a beginning of heterosexual intercourse within the context of an established steady relationship with a partner who is rather older than they are.

Survey research is fairly unanimous in suggesting that sexual intercourse has begun at earlier average ages over recent decades, and that there is more heterosexual activity throughout the teenage years. In general, intercourse begins earlier for young people from lower socio-economic classes and those who have already left education. The range of heterosexual activities for young people appears to be increasing with more of them, for example, reporting oral sex. Increasing rates of heterosexual intercourse are coupled (in UK and elsewhere) with falling rates of births to the under twenties suggesting a wider and more effective use of contraception (Hudson & Ineicher, 1991; Babb, 1993; Forrest, 1990).

Young men

While young men may begin to masturbate at later ages than some young women, the practice is much more common for them. It usually begins around the time of puberty. For the great majority, masturbation to orgasm and ejaculation will long precede any experience of sexual intercourse (see Martinson, 1980). It seems likely that for most, it is through masturbation that boys learn to channel, focus (and name) their sexual feelings and through the accompanying sexual fantasies they learn and organize patterns and definitions of what is sexually arousing for them. In short, it is of central importance for the development of male sexual scripts. Because masturbation is usually a private activity, it does not mean that the scripts do not embody a social and cultural context. The images and models for sexual fantasies will be drawn from the social world boys live in, including the explicitly erotic world of pornography.

When boys begin to make relationships with girls, the cultural expectations that inform them are rather different. For them, sexual desire is to be held to be urgent and uncontrollable, something to be satisfied whenever opportunity allows.

First experiences of sexual intercourse for young men are much more likely than those for young women to take place outside the context of an established relationship and often with a partner with whom they do not have sex again. While reported early experiences of early sexual intercourse is varied, more men than women describe it as being pleasurable. For most men, as with some women, an important psychological process in early experience of intercourse is to bring together the earlier erotic experience of masturbation with that of sex with a partner.

Conclusion

The basic argument outlined is that neither biological accounts of development which privilege physiological development nor surveys of sexual behaviour have proved very illuminating for understanding the development of sexuality. More satisfactory approaches will need to bring in elements of both the biological and social worlds but those that rely on simple notions of socio-biological interaction are also limiting. Attempts have been made to sketch out some of the elements which are needed for more satisfactory approaches. A developmental perspective is required which sees a long-term process beginning early in life and continuing through into adulthood. Puberty, rather than being any kind of starting point for sexual development is, maybe, part of the progress of bodily and social development. The timing of puberty is influenced by the social context of childhood and a child's experience and, in turn, is experienced in different ways depending on the social context. The development of sexuality seems to be very different for girls and women, and men and boys. Each have bodies which are regarded and treated and experienced in different ways according to gender.

A theory of the development of sexuality and sexual behaviour needs as one of its elements a view of the body, not simply as an object, but as a lived-in subject which is perceived by an individual as part of their self. Bodies are also perceived by others in the context of social norms, attitudes and relationships.

Gagnon and Simon's (1973) concept of sexual scripts is endorsed as a useful theoretical way of conceptualizing an individual's pattern of sexual interest, behaviour and desire, including preferences for partners of one or other or both sexes.

Research is still at an early stage. Theoretical ideas require construction hand in hand with empirical studies, which do justice to the complexity of the phenomena to be studied.

Acknowledgements

I would like to thank Ginny Morrow for her comments on an earlier version of this paper. Our research project Transitions to Adulthood is supported by a grant from the Joseph Rowntree Foundation.

References

Amato, P.R. (1993). Children's adjustment to divorce: theories, hypotheses, and empirical support. *Journal of Marriage and the Family*, **55**, 23–38.

Amato, P.R. & Keith, B. (1991). Parental divorce and the well being of children. A meta analysis. *Psychological Bulletin*, **110**, 26–46.

Babb, P. (1993). Teenage conceptions and fertility in England and Wales, 1971–91. *Population Trends*, **74**, 12–15.

Belsky, J., Steinberg, L. & Draper, P. (1991). Childhood experience, interpersonal development, and reproductive strategy: an evolutionary theory of socialisation. *Child Development*, **62**, 647–70.

Brooks-Gunn, J. & Reiter, E.O. (1990). The role of pubertal processes in the early adolescent transition. In *At the Threshold: the Developing Adolescent*, ed. S. Feldman and G. Elliott. Cambridge, Mass: Harvard University Press.

Chilman, C.S. (1986). Some psychosocial aspects of adolescent sexual and contraceptive behavior in a changing American society. In *School-age Pregnancy and Parenthood*, ed. J.B. Lancaster and B.H. Hamburg, pp. 191–218. New York: Aldine de Gruyter.

Chodorow, N. (1978). *The Reproduction of Mothering: Psychoanalysis and the Sociology of Gender*. Berkeley: University of California Press.

Cowie, C. & Lees, S. (1987). Slags and drags. In *Sexuality: A Reader*, ed. *Feminist Review*. London: Virago.

Dombusch, S.M., Carlsmith, J.M., Gross, R.T., Martin, J.A., Jennings, D., Rosenberg, A. & Duke, P. (1981). Sexual development, age, dating: a comparison of biological and social influences upon one set of behaviors. *Child Development*, **52**, 179–85.

Dusek, J.B. (1991). *Adolescent Development and Behavior*. New Jersey: Prentice-Hall.

Elliott, B.J. & Richards, M.P.M. (1991). Children and divorce; educational performance and behaviour before and after parental separation. *International Journal of Law and the Family*, **5**, 258–78.

Eveleth, P.B. (1986). Timing of menarche: secular trend and population difference. In *School-age Pregnancy and Parenthood*, ed. J.B. Lancaster and B.A. Hamburg, pp. 39–52. New York: Aldine de Gruyter.

Eveleth, P.B. & Tanner, J.M. (1976). *Worldwide Variation in Human Growth*. Cambridge: Cambridge University Press.

Fine, M. (1988). Sexuality, schooling and adolescent females: the missing discourse of desire. *Harvard Educational Review*, **58**, 29–53.

Ford, N. (1991). *The Socio-Sexual Lifestyles of Young People in the South West of England*. Exeter, University of Exeter.

Forrest, D.J. (1990). Cultural influences on adolescents' reproductive behaviour. In *Adolescence and Puberty*, eds J. Bancroft & J.M. Reinisch. New York: Oxford University Press.

Furstenberg, F.F., Morgan, S.P., Moore, K.A. and Peterson, J.L. (1987). Race differences in the timing of adolescent intercourse. *American Sociological Review*, **52**, 511–18.

Gaddis, A. & Brooks-Gunn, J. (1985). The male experience of pubertal change. *Journal of Youth and Adolescence*, **14**, 61–9.

Gagnon, J.H. (1983). Age at menarche and sexual conduct in adolescence and young adulthood. In *Menarche, the Transition from Girl to Women*. ed. S.

Golub. Lexington Books.

Gagnon, J.H. & Simon, W. (1973). *Sexual Conduct: the Social Sources of Human Sexuality*. Chicago: Aldine (Hutchison, 1974).

Garn, S.M. (1980). Continuities and change in maturational timing. In *Constancy and Change in Human Development*, ed. O.G. Brim and J. Kagan. Cambridge, Mass.: Harvard University Press.

Gilligan, C. (1982). *In a Different Voice: Psychological Theory and Women's Development*. Cambridge, Mass.: Harvard University Press.

Gilligan, C., Lyons, N. & Hamner (eds) (1989). *Making Connections: The Relational World of Adolescent Girls at Emma Willard School*. Cambridge, Mass.: Harvard University Press.

Gray, R. (1992). Death of the gene: developmental systems strike back. In *Trees of Life*, ed. P. Griffiths. The Hague: Kluwer.

Grief, E.B. & Ulman, K.J. (1982). The psychological impact of menarche on early adolescent females: a review of the literature. *Child Development*, 53, 1413–30.

Griffin, C. (1985). *Typical Girls? Young Women from School to the Job Market*. London: Routledge & Kegan Paul.

Griffin, C. (1993). *Representations of Youth*. Cambridge: Cambridge University Press.

Hill, P. (1993). Recent advances in selected aspects of adolescent development. *Journal of Child Psychology and Psychiatry*, 34, 69–99.

Hirst, J. (1994). Not in front of the grown-ups. A study of the social and sexual lives of 15 and 16 year olds. *Health Research Centre Report No. 6*, Sheffield, Hallam University.

Holland, J., Ramazanoglu, C., Scott, S., Sharpe, S. & Thomson, R. (1990). Sex, gender and power: young women's sexuality in the shadow of AIDS. *Sociology of Health and Illness*, 12, 336–50.

Holland, J., Ramazanoglu, C., Sharpe, S. & Thomson, R. (1991). Pressured pleasure. Young women and the negotiations of sexual boundaries. *WRAP Paper 7*. London: Tuftnell Press.

Hopwood, N.J., Kelch, R.P., Hale, P.M., Mender, T.M., Foster, C.M. & Beitins, I.Z. (1990). The onset of human puberty: biological and environmental factors. In *Adolescence and Puberty*, eds J. Bancroft & J.M. Reinisch. New York: Oxford Univ. Press.

Hudson, F. & Ineicher, B. (1991). *Taking it Lying Down. Sexuality and Teenage Motherhood*. Basingstoke: Macmillan.

Hurlock, B. (1995). *Adolescent Development* (2nd edition). McGraw Hill: New York.

Jackson, S. (1982). *Childhood and Sexuality*. Oxford: Basil Blackwell.

Johnson, A.M., Wadsworth, J., Wellings, R. & Field, J. (1994). *Sexual Attitude and Life Styles*. London: Blackwell Scientific Publications.

Kendall-Tackett, K.A., Williams, L.M. & Finkelhor, D. (1993). Impact of sexual abuse on children: a review and synthesis of recent empirical studies. *Psychological Bulletin*, 113, 164–80.

Kiernan, K.E. (1977). Age at puberty in relation to age at marriage and parenthood: a national longitudinal study. *Annals of Human Biology*, 4, 303–8.

Kiernan, K.E. (1992). The impact of family disruption in childhood on transitions in young adulthood. *Population Studies*, 46, 213–21.

Kinsey, A.C., Pomeroy, W.B. & Martin, C.E. (1948). *Sexual Behavior in the Human*

Male. Philadelphia: Saunders.

Kinsey, A.C., Pomeroy, W.B., Martin, C.E. & Gebhard, P.H. (1953). *Sexual Behavior in the Human Female*. Philadelphia: Saunders.

Knox, E.G., MacArthur, C. & Simons, K.J. (1993). *Sexual Behaviour and Aids in Britain*. London: HMSO.

Kuh, D. & Maclean, M. (1990). Women's childhood experience of parental separation and their subsequent health and socio-economic status in adulthood. *Journal of Biosocial Science*, **22**, 1–15.

Laumann, E.O., Gagnon, J.H., Michael, R.T. & Michaels, S. (1994). *The Social Organisation of Sexuality. Sexual Practices in the United States*. Chicago: University of Chicago Press.

Lerner, R.M., Oulos, J.B. & Knapp, J.R. (1976). Physical attractiveness, physical effectiveness and self-concept in late adolescence. *Adolescence*, **11**, 313–26.

Lewontin, R.C. (1995). *Sex Lies and Social Science*. The New York Review of Books. April, 24–9.

Maccoby, E.E. (1988). Gender as a social category. *Developmental Psychology*, **24**, 755–65.

Maclean, M. & Wadsworth, M.E.J. (1988). The interests of children after parental divorce: a long term perspective. *International Journal of Law and the Family*, **2**, 155–66.

McRobbie, A. (1978). Working class girls and the culture of femininity. In *Women Take Issue*. Women's Studies Group, Centre for Contemporary Cultural Studies. University of Birmingham. London: Hutchinson.

Marshall, W.A. & Tanner, J.M. (1969). Variation in the pattern of pubertal changes in girls. *Archives of Disease in Childhood*, **44**, 291–303.

Marshall, W.A. & Tanner, J.M. (1970). Variation in the pattern of pubertal behaviour in boys. *Archives of Disease in Childhood*, **45**, 13–23.

Martinson, F.M. (1980). Childhood sexuality. In *Handbook of Human Sexuality*, ed. B.B. Wolman and J. Money, pp. 29–59. Englewood Cliffs. NJ: Prentice-Hall.

Meyrick, J. & Harris, R. (1994). Adolescent sexual behaviour, contraceptive use and pregnancy: A review. *ACPP Review and Newsletter*, **16**, 245–51.

Miller, B.C. & Moore, K.A. (1990). Adolescent sexual behavior, pregnancy and parenting: research through the 1980s. *Journal of Marriage and the Family*, **52**, 1025–44.

Miller, B.C. & Heaton, T.B. (1991). Age at first sexual intercourse and the timing of marriage and childbirth. *Journal of Marriage and the Family*, **53**, 719–32.

Miller, B.C., Christophersen, C.R. & King, P.K. (1993). Sexual behavior in adolescence. In *Adolescent Sexuality*, ed. T.P. Gullotta, G.R. Adams and R. Montomayor, Sage: Newbury Park.

Moffitt, T.E., Caspi, A., Belsky, J. & Silva, P.A. (1992). Childhood experience and the onset of menarche: a test of a sociobiological model. *Child Development*, **63**, 47–58.

Monck, E. (1991). Patterns of confiding relationship among adolescent girls. *Journal of Child Psychology and Psychiatry*, **32**, 333–45.

Moore, S. & Rosenthal, D. (1993). *Sexuality in Adolescence*. London: Routledge.

Newcome, S. & Udry, J.R. (1987). Parental marital status effects on adolescent sexual behavior. *Journal of Marriage and the Family*, **49**, 235–40.

Newson, J. & Newson, E. (1963). *Infant Care in an Urban Community*. London:

Allen & Unwin.

Prendergast, S. (1994). The space of childhood: psyche, soma and social existence: menstruation and embodiment at adolescence. In *Debate and Issues in Feminist Research and Pedagogy*, ed. M. Blair and J. Holland, Milton Keynes: Open University Press.

Reiter, E.O. (1986). The neuroendocrine regulation of pubertal onset. In *School-age Pregnancy and Parenthood*, eds J.B. Lancaster and B.A. Hamburg, pp. 53–76. New York: Aldine de Gruyter.

Richards, M.P.M. & Elliott, B.J. (1993). Unpublished observations.

Rodgers, J.L. & Rowe, D.C. (1990). Adolescent sexual activity and mildly deviant behavior. *Journal of Family Issues*, **11**, 274–93.

Rodgers, J.L., Rowe, D.C. & Harris, D.F. (1992). Sibling differences in adolescent sexual behavior: inferring process models from family composition patterns. *Journal of Marriage and the Family*, **54**, 142–52.

Rosenbaum, M. (1979). The changing body image of the adolescent girl. In *Female Adolescent Development*, ed. M. Sugan. Lexington, Mass.: Lexington Books.

Ruble, D.N. & Brooks-Gunn, J. (1982). The experience of menarche. *Child Development*, **53**, 1557–66.

Rust, P.C. (1993). 'Coming out' in the age of social constructionism: sexual identity formation among lesbian and bisexual women. *Gender & Society*, **7**, 50–77.

Sandler, D.P., Wilcox, A.J. & Horney, L.F. (1984). Age at menarche and subsequent reproductive events. *American Journal of Epidemology*, **119**, 765–74.

Schofield, M. (1968). *The Sexual Behaviour of Young People*. Harmondsworth: Penguin Books.

Schofield, M. (1973). *The Sexual Behaviour of Young Adults*. London: Allen Lane.

Simmons, R.G. & Blyth, D.A. (1987). *Moving into Adolescence: the Impact of Pubertal Change and School Context*. New York: Aldine De Gruyter.

Small, S.A. & Lusher, T. (1994). Adolescent sexual activity: an ecological risk-factor approach. *Journal of Marriage and the Family*, **56**, 181–92.

Stattin, H. & Magnusson, D. (1990). *Pubertal Maturation in Female Development*. Hillsdale, New Jersey: Erlbaum.

Steinberg, L. (1988). Reciprocal relation between parent-child distance and pubertal maturation. *Developmental Psychology*, **24**, 122–8.

Steiner-Adair, C. (1990). The body politic. Normal female adolescent development and the development of eating disorders. In *Making Connections. The Relational World of Adolescent Girls at Emma Willard School*, ed. C. Gilligan and N.P. Lyons. Cambridge Mass.: Harvard University Press.

Surbey, M.K. (1990). Family composition, stress and human menarche. In *Socio endocrinology of Primate Reproduction*, ed. T.E. Ziegler and F.B. Bercovitch. New York: Wiley-Liss.

Thompson, S. (1990). Putting a big thing into a little hole: teenage girls' accounts of sexual initiation. *Journal of Sex Research*, **27**, 341–61.

Thomson, R. & Scott, S. (1991). Learning about sex: young women and the social construction of sexual identity. *WRAP Paper No. 4*. London: Tufnell Press.

Thorne, B. (1993). *Gender Play. Girls and Boys in School*. Buckingham: Open University Press.

Tolman, D. (1991). Adolescent girls, women and sexuality: discerning dilemmas of desire. In *Women, Girls and Psychotherapy: Reframing Resistance*, ed. C.

Gilligan, A. Rogers and D. Tolman. New York: Haworth Press.

Tolman, D. (1994). Doing desire. Adolescent girls' struggles for/with sexuality. *Gender and Society*, **8**, 324–42.

Udry, J.R. (1988). Biological predispositions and social control in adolescent sexual behavior. *American Review of Sociology*, **53**, 709–22.

Udry, J.R. (1990). Hormonal and social determinants of adolescent sexual initiation. In *Adolescence and Puberty*, ed. J. Bancroft and J.M. Reinisch. New York, Oxford University Press.

Udry, J.R. & Billy, J.O.G. (1987). Initiation of coitus in early adolescence. *American Review of Sociology*, **52**, 841–55.

Udry, J.R. & Cliquet, R.L. (1982). A cross-cultural examination of the relationship between ages at menarche, marriage and first birth. *Demography*, **19**, 53–63.

Udry, J.R., Talbert, L.M. & Morris, N.M. (1986). Biosocial foundations for adolescent female sexuality. *Demography*, **23**, 217–30.

Ussher, J.M. (1989). *The Psychology of the Female Body*. London: Routledge.

Walster, E., Aronson, V. & Abrahams, D. (1966). Importance of physical attractiveness in dating behavior. *Journal of Personality and Social Psychology*, **4**, 508–16.

Weiss, R. (1975). *Marital Separation*. New York: Basic Books.

West, P., Wight, D. & Macintyre, S. (1996). Heterosexual behaviour of eighteenth year olds in the Glasgow area. *Journal of Adolescence*, in press.

Wight, D. (1990). *The Impact of HIV/AIDS on Young People's Heterosexual Behaviour in Britain: A Literature Review*. Medical Research Council Medical Sociology Research Unit, Glasgow.

13 *Possible relationships between the onset of puberty and female fertility*

LYLIANE ROSETTA

Introduction

Several factors are known to influence the onset of puberty: inherited factors and/or environmental factors like nutrition, infections and strenuous physical activity. Together, they are likely to explain most of the variability in the mean age at puberty among human populations. Unrestricted nutrition is strongly associated with an earlier age of maturation, while a high level of physical training in young girls may be a cause of delayed growth and puberty (Malina, 1983). It has been suggested that there are possible similarities between the process of sexual maturation and the regulation of fertility during adulthood (Dubey *et al.*, 1986). Human fertility is biologically regulated by the hormonal environment of healthy men and women in relationship with their age and lifestyle. Some authors have raised the question of the determinant effect of reproductive impairment during childhood on the level of fertility during adulthood (Kirkwood, Cumming & Aherne, 1987). For example, when young gymnasts or ballet dancers are trained from an age, well before puberty, and show a delay in the onset of menarche, does it mean that their adult reproductive life will be impaired, or altered, even after the end of all training? Furthermore, do girls from traditional populations in the developing world well known for their late onset of puberty relative to European populations, still have low levels of fertility even if they undergo lifestyle changes after migration from rural to more urban settings, with improved nutrition and lower levels of physical activity, and do fertility levels improve in response to improved nutrition (Komlos, 1989)? These issues are examined in this chapter. Recent findings with respect to normal puberty are summarized as are observations on the manipulation of the onset of puberty in different animal species.

206

Normal puberty

In humans, as in other mammals, the maturation of the reproductive system leading to the ability to reproduce involves the hypothalamic–pituitary–gonadal system with its respective hormonal secretions: the gonadotrophin-releasing hormone (GnRH) secreted in a pulsatile pattern from the hypothalamus, and the gonadotrophins, luteinizing hormone (LH) and follicle-stimulating hormone (FSH) secreted in response by the anterior pituitary and mature gametes, and steroid hormones (androgens in males, progestin and oestrogens in females, and inhibins) produced by the gonads. Puberty has been described as 'the transitional stage from childhood to adulthood' (Adams & Steiner, 1988). The process of sexual maturation must be seen as a continuum from fetal life to full maturity, the differentiation between male and female gonads taking place during the fetal life with a final maturation and synchronization undergoing during postnatal development. In human beings, the modifications undergone at hypothalamo-pituitary level to complete a full reproductive axis start some years later during a long and progressive process, well before the first clinical signs of any physical change in secondary sexual characters.

There is still no clear way of identification of the signal which reverses the tendency from a low level of gonadotrophin secreted during mid-childhood (between 2 and 6 years of age) to a slow rise in gonadotrophin secretion which reaches a level high enough to stimulate the secretion of gonadal sex steroids. Indeed, 'the process seems to be a slow, evolutionary one, playing out a series of steps which actually begin *in utero*' (Kulin, 1993). Studies of hormone pulsatility have shown that LH pulse amplitude and frequency increase during puberty in a typical circadian pattern, at first essentially during night-time then during day and night (Kulin, Moore & Santner, 1976; Delemarre-van de Wall *et al.*, 1989; Oerter *et al.*, 1990). Oestradiol concentrations during prepuberty and early puberty in girls show a time-lag pattern of secretion with LH, which may be the time required for the aromatization of oestradiol (E2) synthesis (Goji, 1993).

The development of more sensitive methods such as immunoradiometric assays (IRMA) and immuno-fluorometric assays (IFMA) to trace the activity of the GnRH pulse generator during pubertal transition have shown that even during the prepubertal stage defined by Tanner as Stage I, and which is at about 7 to 8 years of age in girls and about one year later in boys, there is already a baseline activity of the GnRH pulse generator in normal boys and girls, not only during the night but also during the day, although with lower frequency and amplitude (Apter *et al.*, 1993).

A complementary characteristic of the hypothalamic–pituitary–gonadal axis at puberty is the change in the sensitivity of the hypothalamo-pituitary axis to the suppressive effects of gonadal steroids, which seems to be correlated with an enhanced pituitary responsiveness to GnRH (Griffin & Ojeda, 1988).

Delayed puberty

The concept of puberty as a phase in sexual maturation in the continuous transition from fetal life to adulthood has raised in different ways the question of its regulation. Experimental manipulation of the onset of puberty in different animal models as well as clinical and field observation in humans have shed light on the complexity of the system, which involves at least two independent suppressive mechanisms on the hypothalamic–pituitary–gonadal axis before the onset of puberty, one dependent on sex steroids, and another independent of them (Bettendorf & Bettendorf, 1993).

Among independent sex steroid mechanisms, nutrition has been shown in different animal models to play a major role at various stages of reproductive life (I'Anson et al., 1991). In adult female rats, the timing of short-term food restriction at different stages of the oestrous cycle determines the patterns of reduction in mating rates, pregnancy rates or fertility rates, on plasma reproductive hormone concentration or on different aspects of regulation of reproductive function (McClure & Saunders, 1985). For example, in peripubertal female rats, growth and reproductive development are blocked by food restriction (Bronson & Heideman, 1990). In the female lamb, it has been shown that the maintenance of a low body weight by food restriction prevents the normal onset of ovulatory cycling, while the same animals fed ad libitum for several weeks showed rapid catch up in body weight with a concomitant rise in the initiation of ovulatory cycles. In the same study, a control group of chronically food-restricted lambs remained anovulatory. In a separate study, Foster and Olster (1985) demonstrated that nutrition triggers the LH pulse frequency in ovariectomized animals independently of gonadal hormonal feedback.

In juvenile castrated monkeys, dietary restriction leads to a decrease in gonadotropin secretion, in the absence of feedback action by gonadal hormones on LH and FSH (Dubey et al., 1986). By contrast, it has been shown that the prolonged intermittent administration of N-methyl-D-aspartate (NMDA), an analogue of the amino acid aspartate, for 16–30 weeks, to prepubertal monkeys results in the onset of precocious puberty with full activation of the hypothalamic–pituitary–Leydig cells axis and initiation of spermatogenesis (Plant et al., 1989).

In humans, there is some evidence that pubertal development is partly under nutritional control.

Gonadotropin levels were measured among 342 privileged urban children (boys and girls aged 10–14 years) and 347 impoverished rural adolescents from Kenya between 10 and 19 years of age (Kulin *et al.*, 1984). The data were analysed according to age: rural children exhibited a delay in puberty of 3 years in boys and a 2.1 year delay in onset of menarche in girls, compared with privileged urban counterparts of the same ethnic group. Furthermore, an analysis of the same data by stage of sexual maturity showed that later stages of puberty were associated with similar gonadotropin levels in urban and rural populations. This suggests that, even if pubertal maturation is delayed, mature gonadotropin secretion is unchanged when the full maturity of the reproductive axis is reached.

Another source of impairment known to influence the reproductive axis is prolonged exercise, which can interfere with the hypothalamic–pituitary–gonadal axis at the hypothalamic level. It has been shown, in adult female athletes, that endurance training can lead to reproductive dysfunction (Rosetta, 1993), the level of impairment being proportional to the intensity of the exercise above a certain threshold (Goldfarb *et al.*, 1990). Experimental suppression of puberty by prolonged exercise in rats has been demonstrated (Manning & Bronson, 1991): the pulsatile secretions of LH and growth hormone were totally suppressed, relative to controls allowed voluntary exercise. However, mean levels of FSH, prolactin and thyroid stimulating hormone were not affected. Moreover, pulsatile infusion of GnRH re-established puberty in the suppressed animals. The authors of this study suggested a possible additive effect of negative energy balance on hormonal regulation, the animals being fed a diet adequate for normal levels of activity, but probably not enough to cover the needs for maintenance of energy balance during high levels of physical activity.

In girls, the role of high levels of training prior to puberty in the regulation of ovulatory function have been questioned, since growth and puberty in female gymnasts both tend to be delayed (Mansfield & Emans, 1993), while the incidence of menstrual dysfunction is higher than in girls having initiated training soon after menarche (Baer, 1993). A prospective study was carried out among two groups of adolescent female athletes to assess the effect of training on their physical and sexual maturation. Twenty-two adolescent gymnasts and 21 swimmers were examined longitudinally for 2 to 4 years by Theintz *et al.* (1993), this study providing evidence of a reduction of growth rates with extremely intensive training: a decrease in predicted mean height with time was found in the gymnasts but not the swimmers. In addition, during the study period, the proportion of athletes having had menarche increased from 4.5% to 50% for gymnasts and from 28.6% to 71.4% for swimmers. The menarcheal age was significantly delayed in gymnasts (14.5 ± 1.2 years; $n = 11$) compared to the group of swimmers (12.9±0.9 years; $n=15$). An unexpectedly low incidence of menstrual dysfunction was found in a large group of Danish female elite swimmers performing hard physical endurance training, given that at similar intensity and duration of training, long-distance runners were known for their high level of menstrual dysfunction (Faunø, Kålund & Kanstrup, 1991). Different partitioning of macronutrients leading to different body composition, with higher body mass and higher percentage of subcutaneous fat for swimmers compared to runners, has been proposed by way of explanation.

Even if nutrition and prolonged exercise emerge as serious candidates of non-sex-steroid-dependent factors of regulation of the onset of puberty, there is possible complementarity or interference by energy expenditure, energy balance, stress or any physiological signals they might produce. Two major hypotheses have been proposed to examine the nature of the signal for the onset of puberty. The first is the 'body weight' hypothesis which suggests that there may be a critical weight, weight change or body fatness determinant in the process of sexual maturation (Frisch, 1988). The second is the 'metabolic hypothesis' which proposes that blood-borne metabolic cues which can be metabolites, hormones or a combination of both, and which inform the hypothalamic-pituitary axis about metabolic readiness for puberty. Recently, Suttie et al. (1991) used a lamb animal model to test both hypotheses, and showed that body weight was not a determinant of puberty onset but associated with metabolic cues. After a study of the relationship between body fat mass, body fat distribution and pubertal development in girls, de Ridder et al. (1992) came to the conclusion that 'neither body fat mass nor body fat distribution seem to trigger puberty onset in healthy girls'. In rats and monkeys, the 'body weight' hypothesis has also been rejected (Glass, Herbert & Anderson, 1986; Bronson & Manning, 1991).

Evidence of subfertility in adulthood or seasonal decrease in fertility in relation to lifestyle

Extensive observational surveys in various human populations have shown, during the last ten years, that there is significant interpopulation variability in ovarian function among adult women. Even immediately after puberty, a comparison of the progesterone activity in adolescents living in Britain and Thailand, respectively, have shown significantly greater progesterone activity in British girls compared to their Thai counterparts when matched for chronological and gynaecological ages, mainly in older girls (Danutra et al., 1989). This result can be compared with those of Ellison et al. (1993) for adult women with regular menstrual cycles from different countries or socio-economic levels. Women from rural Zaire, Nepal or Poland, when compared to well-off American controls, exhibit lower levels of plasma gonadal hormones with evidence of high incidence of anovulatory cycles and frequently inadequate luteal phases.

Similar findings have been published from a large epidemiological survey carried out among 3250 rural Chinese women recruited in 65 counties in China and compared with British controls (Key *et al.*,1990); the mean concentration of oestradiol was 36% higher in the group of 35–44 year-old British controls than in the same age-group of Chinese women. The difference in oestradiol or oestrone concentration seems to be positively correlated with the level of fat intake and negatively correlated with the quantity of dietary fibre. Experimental manipulation of the fat composition of the diet of healthy premenopausal women showed a significant increase in follicular phase duration in women consuming a low fat diet (Reichman *et al.*, 1992). Furthermore, Asian populations eating a traditional, mostly vegetarian diet, have significantly lower levels of plasma oestrone, oestradiol, testosterone and androstenedione than omnivorous Caucasian controls (Goldin *et al.*, 1986). In addition, the ratio of urinary-to-faecal oestrogen excretion was higher for oestriol, oestrone and oestradiol measured in the group of Asian women compared to their western counterparts.

Studies of risk factors in the aetiology of breast cancer have suggested that a vegetarian diet may influence the gonadal steroid metabolism (Armstrong & Doll, 1975). To test this hypothesis, Hill and coworkers (1984) have compared the duration of the menstrual cycle after a change in the usual diet with and without meat among omnivorous Caucasian women and vegetarian South African women. The vegetarian diet was associated with a shortening of the follicular phase during the second menstrual cycle among Caucasian women, with a decrease in the pulsatile release of LH and FSH, and with a decrease of the pituitary response to GnRH stimulation. Among the South African women, transfer to a western diet for a two-month period resulted in an increase in the duration of the follicular phase. In both groups, body weight was not modified at the end of the experiment, their diet having been carefully established to match the usual energy intake in each group. It is interesting to note that, after food manipulation in healthy adult women, the modification of their menstrual status was restored soon after a return to their usual diet.

In addition, Pedersen *et al.* (1991) comparing traditional vegetarian women with non-vegetarian premenopausal healthy women, found a significantly higher percentage of menstrual irregularities in vegetarians compared with non-vegetarians. Significantly longer duration of menstrual cycles have been observed in women in Highland Papua New Guinea relative to Western women; this might be partly attributed to dietary differences (Johnson *et al.*, 1987).

The adequacy of food intake and experimental food restriction have both been shown to play a determinant role in the regulation of reproductive function at the hypothalamic level. Food restriction and subsequent re-feeding modulates gonadotrophin secretion and serum testosterone levels in male rats (Howland, 1975). More recently, similar effects of dietary restriction have been investigated in female lambs, which became hypogonadotropic after having been chronically undernourished. These animals showed delayed puberty due to a suppression of LH pulse frequency (Foster *et al.*, 1989), whereas unrestricted re-feeding of chronically food-restricted lambs resulted in significantly increased synthesis, storage and secretion of LH, FSH and GH (Landefeld *et al.*, 1989). Similar types of food restriction on the regulation of reproductive function have also been investigated in primates and humans (Cameron, Helmreich & Schreihofer, 1993; Rödjmark, 1987; Loucks *et al.*, 1994).

It has been shown that short-term food shortage directly interferes with the central drive of the reproductive axis in primates and humans in slowing the pulsatility of the GnRH pulse generator, which is mirrored by a reduction in LH pulse frequency but not amplitude, and a concomitant decrease in the pulsatile secretion of testosterone in adult male Rhesus monkeys after one day of fasting (Cameron & Nosbisch, 1991), and in normal male volunteers after 48 hours of fasting (Cameron *et al.*, 1991). The restoration of LH pulsatility in male rhesus monkeys after a day of fasting was directly proportional to the size of the re-feed meal (Parfitt, Church & Cameron, 1991). The activation of the adrenal axis following a meal missed is not the causal agent in the disruption of the GnRH pulse generator (Helmreich, Mattern & Cameron, 1993). To discriminate between stress and metabolic signals involved in this regulation, an intragastric nutrient infusion was used to re-feed the animals and authors showed that the signal is associated with nutrient intake (Schreihofer, Parfitt & Cameron, 1993*a*; Schreihofer *et al.*, 1993*b*). Furthermore, the inhibitory effect of short-term fasting on the hypothalamic and/or pituitary stimulation of the Leydig cell function can be blocked in normal healthy men by oral glucose supplementation (Rödjmark, 1987).

Discussion

There appears to be total reversibility of menstrual impairment in adults when due to the consumption of a vegetarian diet or endurance training. Indeed, female long-distance runners who give up training for various causes, but mainly when they are physically injured, return to a more normal menstrual cycle after a few months. The delay in the return to fertility is associated with the extent of impairment and the duration of exposure to the stressor leading to this impairment. There are various examples of female marathon runners who have been amenorrheic for several years without any pathological cause. When they have had to give up training for reasons such as chronic tendinitis, their menses usually return after 2 or 3 months of complete rest. In this case there are two factors conjoined: (i) the cessation of the daily source of impairment, in terms of training sessions; and (ii) a delay in recovery of normal menstrual function, probably linked to profound modifications of the ultrastructure of GnRH neurons after a long period of hormonal deprivation. A dramatic decrease of synaptic input and neuronal activity has been observed in female rhesus monkeys after ovariectomy, which was partially restored after hormonal replacement (Witkin *et al.*, 1991).

Further evidence in favour of reversibility is that, in developing countries, there is a clear trend toward shorter duration of postpartum amenorrhea and earlier return of ovulatory cycles among lactating women even in moderately poor socio-economic families (Dada, 1992). The secular trend toward a shortening of the postpartum amenorrhea duration shown in Norway can be related to the same process (Liestøl, Rosenberg & Walløe, 1988). Even if a part of the dramatic decrease observed in the early 1900s was probably due to improved hygiene and modifications in breast-feeding patterns, these factors are linked to a concomitant increase in socio-economic level and improved nutrition.

Conversely, it has been recently suggested that there may be several critical periods of sensitivity to the organizational effects of androgens during gestation and between birth and puberty, and there is some evidence from animal experimentation that some of them are not reversible (Mann *et al.*, 1993). In humans, a follow-up study of growth and development carried out among 216 adolescent Barbadians who have histories of moderate to severe protein-energy malnutrition in their first year of life, have shown delays in physical growth and sexual maturation, especially in girls (Galler, Ramsey & Solimano, 1985). There is clinical evidence from children adopted into privileged conditions that early puberty is likely to occur among children having had serious episodes of malnutrition during their childhood, when they are transferred into well-off families. Indian girls adopted in Sweden have shown a significantly early menarche (median age at puberty = 11.5 years) compared either to Indian privileged girls (12.4–12.9 years) or to Indian girls from rural areas raised in India (13.7 years), or to Swedish girls born and raised in Sweden (13.0 years). The date of arrival in the country of adoption is correlated with the age at menarche.

Adoption after 3 years of age is associated with a higher risk of very early sexual maturation (Proos, Hofvander & Tuvemo, 1991), and in turn, earlier menarche is associated with short final stature (Proos, 1993). The same findings have been reported for children with central precocious puberty adopted in Belgium from developing countries. To test the hypothesis of critical period of refeeding after nutritional deprivation on the timing of hypothalamic maturation, authors used a male rat model. An accelerated hypothalamic and testicular maturation was observed in animals with small body size, and the timing of hypothalamic maturation was affected by changes in nutritional conditions when the refeeding occurred before weaning (Bourguignon et al., 1992). In children, the same acceleration of hypothalamic and sexual maturation, accompanied by a premature growth pattern leads to short final stature and possible subsequent difficulties in reproduction for women of very small body size, with greater risk of complications in delivery. Nevertheless, this does not suggest that these children are likely to be less fertile than normal maturers.

It seems likely that there is some critical period of development during which an impairment of reproductive function might be irreversible. Given that there is a continuum in the maturation of the hypothalamic–pituitary–gonadal axis, it is possible that malnutrition and/or any cause of disruption of the activity of the hypothalamic–pituitary–gonadal axis during early postnatal life could have some impact on reproductive life (Mann et al., 1993). In addition to individual variability in fertility, it seems that early environment may be important for the development of mature levels of fertility. Dietary deficiencies during fetal life or the postnatal period may have definitive consequences on reproductive potential, but there is no evidence for a lack of reversibility associated with later impairment at present. This has to be tested, possibly by the study of migrant populations, comparing adult native fertility with adult migrant fertility among subjects born and raised in the same country during childhood.

References

Adams, L.A. & Steiner, R.A. (1988). Puberty. Oxford Reviews of Reproductive Biology, 10, 1–52.

Apter, D., Bützow, T.L., Laughlin, G.A. & Yen, S.S.C. (1993). Gonadotropin-releasing hormone pulse generator activity during pubertal transition in girls: pulsatile and diurnal patterns of circulating gonadotropins. Journal of Clinical Endocrinology and Metabolism, 76, 940–9.

Armstrong, B. & Doll, R. (1975). Environmental factors and cancer incidence and mortality in different countries with special reference to dietary practices. International Journal of Cancer, 15, 617–31.

Baer, J.T. (1993). Endocrine parameters in amenorrheic and eumenorrheic adolescent female runners. *International Journal of Sports and Medicine*, **14**, 191–5.

Bettendorf, M. & Bettendorf, G. (1993). Search for a biological clock in the ontogeny of puberty. *Human Reproduction*, **8**, 791–2.

Bourguignon, J-P., Gérard, A., Gonzalez, M-L, A., Fawe, L. & Franchimont, P. (1992). Effects of changes in nutritional conditions on timing of puberty. Clinical evidence from adopted-children and experimental studies in male-rat. *Hormone Research*, **38**, 97–105.

Bronson, F.H. & Heideman, P.D. (1990). Short-term hormonal responses to food intake in peripubertal female rats. *American Journal of Physiology*, **259**, R25–R31.

Bronson, F.H. & Manning, J.M. (1991). The energetic regulation of ovulation: a realistic role for body fat. *Biology of Reproduction*, **44**, 945–50.

Cameron, J.L. & Nosbisch, C. (1991). Suppression of pulsatile luteinizing-hormone and testosterone secretion during short-term food restriction in the adult male Rhesus-monkey (*Macaca mulatta*). *Endocrinology*, **128**, 1532–40.

Cameron, J.L., Weltzin, T.E., McConaha, C., Helmreich, D.L. & Kaye, W.H. (1991). Slowing of pulsatile luteinizing-hormone secretion in men after 48 hours of fasting. *Journal of Clinical Endocrinology and Metabolism*, **73**, 35–41.

Cameron, J.L., Helmreich, D.L. & Schreihofer, D.A. (1993). Modulation of reproductive hormone secretion by nutritional intake: stress signals versus metabolic signals. *Human Reproduction*, **8**, 162–7.

Dada, O.A. (1992). Brief description of WHO protocol for data collection. *Journal of Biosocial Science*, **24**, 379–81.

Danutra, V., Turkes, A., Read, G.F., Wilson, D.W., Griffiths, V., Jones, R. & Griffiths, K. (1989). Progesterone concentrations in samples of saliva from adolescent girls living in Britain and Thailand, two countries where women are at widely different risk of breast cancer, *Journal of Endocrinology*, **121**, 375–81.

Delemarre-van de Wall, H.A., Plant, T.M., van Rees, G.P. & Shoemaker, J., eds (1989). *Control of the Onset of Puberty III*. Amsterdam: Excerpta Medica.

Dubey, A.K., Cameron, J.L., Steiner, R.A. & Plant, T.M. (1986). Inhibition of gonadotropin secretion in castrated male rhesus monkeys (*Macaca mulatta*) induced by dietary restriction: analogy with the prepubertal hiatus of gonadotropin release. *Endocrinology*, **118**, 518–25.

Ellison, P.T., Lipson, S.F., O'Rourke, M.T., Bentley, G.R., Harrigan, A.M., Panter-Brick, C. & Vitzhum, V.J. (1993). Population variation in ovarian function. *Lancet*, **342**, 433–4.

Faunø, P., Kålund, S. & Kanstrup, I.-L. (1991). Menstrual patterns in Danish elite swimmers. *European Journal of Applied Physiology*, **62**, 36–9.

Foster, D.L. & Olster, D.H. (1985). Effect of restricted nutrition on puberty in the lamb: patterns of tonic luteinizing hormone (LH) secretion and competency of the LH surge system. *Endocrinology*, **116**(1), 375–81.

Foster, D.L., Ebling, F.J.P., Micha, A.F., Vannerson, I.A., Bucholtz, D.C., Wood, R.I., Suttie, J.M. & Fenner, D.E. (1989). Metabolic interfaces between growth and reproduction. I. Nutritional modulation of gonadotropins, prolactin, and growth hormone secretion in the growth-limited female lamb. *Endocrinology*, **125**, 342–50.

Frisch, R.E. (1988). Fatness and fertility. *Scientific American*, **258**, 88–95.

Galler, J.R., Ramsey, F. & Solimano, G. (1985). A follow-up study of the effects of

216   L. Rosetta

early malnutrition on subsequent development. I. Physical growth and sexual maturation during adolescence. *Pediatric Research*, **19**, 518–23.

Glass, A.R., Herbert, D.C. & Anderson, J. (1986). Fertility onset, spermatogenesis, and pubertal development in male rats: effects of graded underfeeding. *Pediatric Research*, **20**, 1161–7.

Goji, K. (1993). Twenty-four-hour concentration profiles of gonadotropin and estradiol (E_2) in prepubertal and early pubertal girls: the diurnal rise of E_2 is opposite the nocturnal rise of gonadotropin. *Journal of Clinical Endocrinology and Metabolism*, **77**, 1629–35.

Goldfarb, A.H., Hatfield, B.D., Armstrong, D. & Potts, J. (1990). Plasma beta-endorphin concentration: response to intensity and duration of exercise. *Medicine and Science in Sports and Exercise*, **22**, 241–4.

Goldin, B.R., Adlercreutz, H., Gorbach, S.L., Woods, M.N., Dwyer, J.T., Conlon, T., Bohn, E. & Gershoff, S.N. (1986). The relationship between estrogen levels and diets of Caucasian American and Oriental immigrant women. *American Journal of Clinical Nutrition*, **44**, 945–53.

Griffin, J.E. & Ojeda, S.R. (1988). *Textbook of Endocrine Physiology*. New York, Oxford: Oxford University Press.

Helmreich, D.L., Mattern, L.G. & Cameron, J.L. (1993). Lack of a role of the hypothalamic-pituitary-adrenal axis in the fasting-induced suppression of luteinizing hormone secretion in adult male rhesus monkeys (*Macaca mulatta*). *Endocrinology*, **132**, 2427–37.

Hill, P., Garbaczewski, L., Haley, N. & Wynder, E.L. (1984). Diet and follicular development, *American Journal of Clinical Nutrition*, **39**, 771–7.

Howland, B.E. (1975). The influence of feed restriction and subsequent re-feeding on gonadotrophin secretion and testosterone levels in male rats. *Journal of Reproduction and Fertility*, **44**, 429–36.

I'Anson, H., Foster, D.L., Foxcroft, G.R. & Booth, P.J. (1991). Nutrition and reproduction. In *Oxford Reviews of Reproductive Biology*, ed. S.R. Milligan, Vol. 13. Oxford: Oxford University Press.

Johnson, P.L., Wood, J.W., Campbell, K.L. & Maslar, I.A. (1987). Long ovarian cycles in women in highland New Guinea. *Human Biology*, **59**, 837–45.

Key, T.J.A., Chen, J., Wang, D.Y., Pike, M.C. & Boreham, J. (1990). Sex hormones in women in rural China and in Britain. *British Journal of Cancer*, **62**, 631–6.

Kirkwood, R.N., Cumming, D.C. & Aherne, F.X. (1987). Nutrition and puberty in the female. *Proceedings of the Nutrition Society*, **46**, 177–92.

Komlos, J. (1989). The age at menarche and age at first birth in an undernourished population. *Annals of Human Biology*, **16**, 463–6.

Kulin, H.E., Moore, R.G. & Santner, S.J. (1976). Circadian rhythms in gonadotropin excretion in prepubertal and pubertal children. *Journal of Clinical Endocrinology and Metabolism*, **42**, 770–3.

Kulin, H.E., Bwibo, N., Mutie, D. & Santner, S.J. (1984). Gonadotropin excretion during puberty in malnourished children. *Journal of Pediatrics*, **105**, 325–8.

Kulin, H.E. (1993). Editorial: Puberty: when? *Journal of Clinical Endocrinology and Metabolism*, **76**, 24–5.

Landefeld, T.D., Ebling, F.J.P., Suttie, J.M., Vannerson, L.A., Padmanabhan, V. Beitins, I.Z. & Foster, D.L. (1989). Metabolic interfaces between growth and reproduction. II. Characterization of changes in messenger ribonucleic acid concentrations of gonadotrophin subunits, growth hormone and prolactin in nutritionally growth limited lambs and the differential effects of increased nutrition. *Endocrinology*, **125**, 351–6.

Liestøl, K., Rosenberg, M. & Walløe, L. (1988). Lactation and post-partum amenorrhea: a study based on data from three Norwegian cities 1860–1964. *Journal of Biosocial Science*, **20**, 423–34.

Loucks, A.B., Heath, E.M., Verdun, M. & Watts, J.R. (1994). Dietary restriction reduces luteinizing hormone (LH) pulse frequency during waking hours and increases LH pulse amplitude during sleep in young menstruating women. *Journal of Clinical Endocrinology and Metabolism*, **78**, 910–15.

McClure, T.J. & Saunders, J. (1985). Effects of withholding food for 0–72 h on mating, pregnancy rate and pituitary function in female rats. *Journal of Reproduction and Fertility*, **74**, 57–64.

Malina, R.M. (1983). Menarche in athletes: a synthesis and hypothesis. *Annals of Human Biology*, **10**, 1–24.

Mann, D.R., Akinbami, M.A., Gould, K.G., Tanner, J.M. & Wallen, K. (1993). Neonatal treatment of male monkeys with a gonadotropin-releasing hormone agonist alters differentiation of central nervous system centers that regulate sexual and skeletal development. *Journal of Clinical Endocrinology and Metabolism*, **76**, 1319–24.

Manning, J.M. & Bronson, F.H. (1991). Suppression of puberty in rats by exercise: effects on hormone levels and reversal with GnRH infusion. *American Journal of Physiology*, **260**, R717–23.

Mansfield, M.J. & Emans, S.J. (1993). Growth in female gymnasts. Should training decrease during puberty? *Journal of Pediatrics*, **122**, 237–40.

Oerter, K.E., Uriarte, M.M., Rose, S.R., Barnes, K.M. & Cutler, G.B. (1990). Gonadotropin secretory dynamics during puberty in normal girls and boys. *Journal of Clinical Endocrinology and Metabolism*, **71**, 1251–8.

Parfitt, D.B., Church, K.R. & Cameron, J.L. (1991). Restoration of pulsatile luteinizing hormone secretion after fasting in rhesus monkeys (*Macaca mulatta*): dependence on size of the refeed meal. *Endocrinology*, **129**, 749–56.

Pedersen, A.B., Bartholomew, M.J., Dolence, L.A., Aljadir, L.P., Netteburgh, K.L. & Lloyd, T. (1991). Menstrual differences due to vegetarian and nonvegetarian diets. *American Journal of Clinical Nutrition*, **53**, 879–85.

Plant, T.M., Gay, V.L., Marshall, G.R. & Arslan, M. (1989). Puberty in monkeys is triggered by chemical stimulation of the hypothalamus. *Proceedings of the National Academy of Sciences of the United States*, **86**, 2506–10.

Proos, L.A. (1993). Anthropometry in adolescence. Secular trends, adoption, ethnic and environmental differences. *Hormone Research*, **39**, 18–24.

Proos, L.A., Hofvander, Y. & Tuvemo, T. (1991). Menarcheal age and growth pattern of Indian girls adopted in Sweden. *Acta Paediatrica Scandinavica*, **80**, 852–8.

Ridder, C.M. de, Thijssen, J.H.H., Bruning, P.F., Van den Brande, J.L., Zonderland, M.L. & Erich, W.B.M. (1992). Body fat mass, body fat distribution, and pubertal development: a longitudinal study of physical and hormonal sexual maturation of girls. *Journal of Clinical Endocrinology and Metabolism*, **75**, 442–6.

Reichman, M.E., Judd, J.T., Taylor, P.R., Nair, P.P., Jones, D.Y. & Campbell, W.S. (1992). Effect of dietary fat on length of the follicular phase of the menstrual cycle in a controlled diet setting. *Journal of Clinical Endocrinology* and Metabolism, **74**, 1171–5.

Röjdmark, S. (1987). Influence of short-term fasting on the pituitary–testicular axis in normal men. *Hormone Research*, **25**, 140–6.

Rosetta, L. (1993). Female reproductive dysfunction and intense physical training. In *Oxford Reviews of Reproductive Biology*, Vol. **15**, edited by S.R. Milligan, pp. 113–41. Oxford: Oxford University Press.

Schreihofer, D.A., Parfitt, D.B. & Cameron, J.L. (1993a). Suppression of luteinizing- hormone secretion during short term fasting in male rhesus monkeys. The role of metabolic versus stress signals. *Endocrinology*, **132**, 1881–9.

Schreihofer, D.A., Amico, J.A. & Cameron, J.L. (1993b). Reversal of fasting-induced suppression of luteinizing-hormone (LH) secretion in male rhesus monkeys by intragastric nutrient infusion. Evidence for rapid stimulation of LH by nutritional signals. *Endocrinology*, **132**, 1890–7.

Suttie, J.M., Foster, D.L., Veenvliet, B.A., Manley, T.R. & Corson, I.D. (1991). Influence of food intake but independence of body weight on puberty in female sheep, *Journal of Reproduction & Fertility*, **92**, 33–9.

Theintz, G.E., Howald, H., Weiss, U. & Sizonenko, P.C. (1993). Evidence for a reduction of the growth-potential in adolescent female gymnasts. *Journal of Pediatrics*, **122**, 306–13.

Witkin, J.W., Ferin, M., Popilskis, S.J. & Silvermann, A-J. (1991). Effects of gonadal steroids on the ultrastructure of GnRH neurons in the Rhesus monkey: synaptic input and glial apposition. *Endocrinology*, **129**, 1083–92.

14 Early environment, long latency and slow progression of late onset neuro-degenerative disorders

RALPH M. GARRUTO

Introduction

The effects of early life events on the outcome of adult onset disease is a research area still in its infancy. Although there is considerable anecdotal evidence as well as some empirical observations of the timing of exposure to infectious agents in childhood and later outcome of adult onset disorders, few systematic studies have been pursued longitudinally, particularly with regard to chronic, non-infectious multifactorial adult diseases. This is certainly true for the major chronic late onset neuro-degenerative disorders of humankind that include Alzheimer's disease, parkinsonism, amyotrophic lateral sclerosis (ALS) and multiple sclerosis. Many of these diseases are now thought not to be single entity disorders, but to contain subtypes or subpopulations of patients whose neurological conditions are caused by different insults that are expressed through a common pathway leading to similar neuro-degenerative changes over time.

The nervous system has a limited number of responses to various types of insults, be they genetic or non-genetic, and it can be speculated that neuronal degeneration leading to eventual onset of clinical disease is governed by a process of exposure to long-term insults which give rise to adult disease. Age, time and the sequence of events appear to be the most important factors giving rise to late onset neuro-degenerative diseases. At some point during adulthood normal neuronal abiotrophy (slow neuronal loss through time) occurs in conjunction with a pathological loss of neurons that can reach 80% or 90% in some brain areas before clinical manifestations are seen (Strong & Garruto, 1994). Exactly how this takes place is unknown, but the kind and sequence of events are worthy of research using methodological designs that are initiated early in life, *in utero*, during infancy, childhood, adolescence, but long before onset of

219

clinical disease. Such studies represent high risk research and it may be difficult to obtain funding support for them.

In this chapter current ideas on diseases of long-latency and slow progression will be presented along with new thinking on age-related disease continua, slow toxins, and of the process of protein fibrillization and molecular reorganization that likely have a long temporal sequence and lead to neuronal degeneration. A discussion of these ideas and processes are couched within a framework of high incidence neuro-degenerative diseases, ALS and parkinsonism-dementia (PD) occurring in the western Pacific, and their relationship to Alzheimer's disease. Additionally, experimental attempts to establish *in vivo* and *in vitro* models of neuronal degeneration and neurofibrillary tangle formation, the hallmark neuropathology of these disorders, are described and related back to early life events as risk factors for adult onset disease.

Diseases of long latency and slow progression

The major adult-onset neuro-degenerative disorders, including ALS, parkinsonism and Alzheimer's disease, are all diseases of late onset with slow progression of clinical symptoms, culminating uniformly in death. Neuropathologically, they all have in common fundamental abnormalities in biosynthesis, phosphorylation and/or catabolism of proteins that coalesce or aggregate both intra- and extraneuronally in the accumulation of cytoskeletal proteins, amyloid β-protein and other fibrillary proteins, accompanied by severe neuronal loss and atrophy in disease specific subanatomic regions of the brain and/or spinal cord. A second group of human disorders, kuru, Creutzfeldt–Jakob disease, fatal familial insomnia, and the Gerstmann–Sträussler–Scheinker syndrome, termed the infectious brain amyloidoses, represent a long subclinical (latent) stage where fibrillization and configurational change in proteins occur (Gajdusek, 1994). Thus, both groups of disorders are relentlessly progressive and uniformly fatal and form a biological continuum linked in their pathogenic process by a common pathway and/or initiating insult(s).

It has been postulated that the degenerative process of these disorders may be a simple augmentation of the normal abiotrophy or attrition of neurons during ageing (Eisen & Calne, 1990). While an attractive hypothesis, the rates of loss of neurons are precipitous and exceed by severalfold the naturally occurring rates. Thus, the normal process of abiotrophy is accompanied by a very active process, the triggering of a cascade of neuronal loss which coincides with clinical onset and progression of the course of the disease. What triggers this accelerated neuronal death is not known nor are the mechanisms of targeting of specific neuronal populations. Moreover, these disorders are age-related phenomena with progressively increasing incidence rates beginning around the sixth decade of life, with normal neuronal estimates in healthy individuals remaining essentially stable until this time, and declining thereafter.

Many late onset disorders with long latency and slow progression are likely to represent disease syndromes rather than single disease entities. This should be expected, given the limited phenotypic repertoire of the adult nervous system. Part of the difficulty in understanding the pathogenesis of these syndromes lies in the utilization of categorical clinical criteria to define continuous processes, that while superficially similar, may reflect varied pathophysiological processes. However these diseases are ultimately classified, it is obvious that the process of neuronal degeneration is one with a long temporal sequence, with clinical recognition of these disorders representing the end stage of this process. If the disease process is to be truly understood, an appropriate methodology must be used which allows evaluation of the total life course of the individual and the development of concepts of accelerated neuronal abiotrophy, neuronal cascading and altered disease 'set points' (Brant & Pearson, 1994; Strong & Garruto, 1994).

Neuro-degenerative disorders of the western Pacific are associated with a long temporal sequence

Perhaps two of the best examples of disorders with long temporal sequences that give some insight into the importance of life history events and the ability to test a series of environmental and genetic hypotheses regarding these events are ALS and PD of the western Pacific (Figure 14.1). During the past several decades, high-incidence foci of ALS and PD have been studied in non-western anthropological populations in the Pacific Basin aimed at identifying their aetiology and mechanisms of pathogenesis and their relationship to other neuro-degenerative disorders, such as Alzheimer's disease, parkinsonism and early neuronal ageing (Garruto, 1991). Pacific ALS and PD are two progressive and fatal neurological disorders that occur hyperendemically in different cultures, in different ecological zones and among genetically divergent populations in the Mariana Islands (Garruto, 1991), the Kii Peninsula of Japan (Yase, 1980; Shiraki & Yase, 1991) and southern West New Guinea (Gajdusek, 1963; Gajdusek & Salazar, 1982).

The cross-disciplinary approach to these intriguing neurological disorders and the accumulated epidemiological and cellular and molecular evidence strongly implicate environmental factors in their causation, specifically the role of aluminium and its interaction with calcium. The

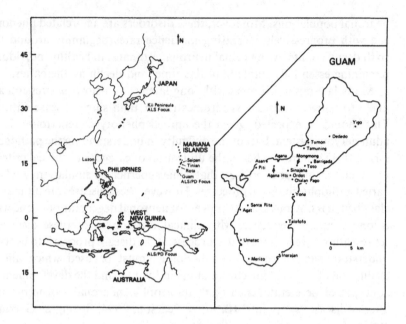

Figure 14.1. Map of the western Pacific showing the geographic location of the three high incidence foci of amyotrophic lateral sclerosis (ALS) and parkinsonism-dementia (PD). The detailed insert map of Guam shows the geographic location of all villages on the island at the height of the epidemic.

dramatic disappearance over the past 30 years of these disorders, with original incidence rates 50–150 times higher than elsewhere in the world, clearly supports major environmental factors in the disease process, coincident with decreased isolation, increased contact and increased westernization of these populations. A clearer understanding of these natural paradigms has had, and will continue to have, far-reaching implications for the study of Alzheimer's disease and other disorders with long latency and slow progression. Although the primary emphasis here is on the Guamanian focus of ALS and PD, which is the best studied of the three high incidence Pacific foci, some aspects and cross-comparisons of these disorders in West New Guinea and the Kii Peninsula of Japan are also described, as is their neuropathological link to Alzheimer's disease. As an outgrowth of these human orientated studies, some experimental evidence from studies of *in vivo* and *in vitro* models of neuronal degeneration and the available experimental evidence implicating early life events in the disease process will also be summarized.

Clinical features and neuropathological changes

In 1953, Arnold and his associates reported an unusually high incidence of ALS among the indigenous Chamorro population of Guam (Arnold, Edgren & Palladino, 1953). Subsequent reports by Tillema & Wijnberg (1953) and Kurland & Mulder (1954, 1955) verified and expanded Arnold's findings. Mulder, Kurland & Iriarte (1954) also reported 22 patients with parkinsonism, some of whom had dementia. This combination of parkinsonism with dementia was subsequently studied in detail by Hirano and colleagues (1961*a,b*), who concluded that it represented a unique disease, which they called parkinsonism-dementia (PD). It was subsequently shown that both ALS and PD occurred in numerous Guamanian families and that a number of patients had clinical and neuropathological changes of both diseases (Kurland & Mulder, 1955; Elizan *et al.*, 1966; Chen, 1979; Plato *et al.*, 1967, 1986; Bailey-Wilson *et al.*, 1993). The initial findings provided the impetus for the National Institute of Neurological Diseases and Stroke in 1956 to establish the NINDS Guam Research Center. Scientists at the Center and their colleagues on the mainland embarked on an intensive multidisciplinary research effort towards understanding the aetiology of ALS and PD, with an ultimate goal of treatment and prevention of disorders which had plagued the Chamorro people for at least a century.

ALS on Guam, known locally as *lytico* or *paralytico*, is a neuromuscular disorder which affects the upper and lower motor neurons of the brain and spinal cord. Clinically, the disease is indistinguishable from classical sporadic ALS worldwide. It is characterized by progressive muscle weakness, atrophy, fasciculations (muscle jumping) and spasticity, leading to paralysis and eventually death within four years (Table 14.1). Neuropathologically, Guamanian ALS differs from classical sporadic ALS found worldwide in that the loss of neurons in the brain and spinal cord are also characterized by extensive development of neurofibrillary tangles (NFTs) without associated senile plaques (Table 14.2) (Garruto, 1991). The NFTs seen in ALS and PD appear at the cellular and molecular level to be identical to those found in brains of Alzheimer disease patients (Guiroy *et al.*, 1987, 1993; Shankar *et al.*, 1989).

PD, known locally as *bodig* or *rayput*, occurs in high incidence in the same villages and the same sibships as ALS, and occasionally a patient may have both disorders (Table 14.3). PD is an extrapyramidal disease characterized clinically by parkinsonian features, including staring gaze, muscular rigidity, tremor, bradykinesia (slow voluntary movement) and a shuffling gait. These symptoms are combined with onset of a progressive dementia early in the disease course (Table 14.1). Neuropathologically,

Table 14.1. *Comparison of clinical features of high incidence ALS and PD in the Western Pacific*

Clinical features	Guam ALS	Guam PD	West New Guinea ALS	West New Guinea PD	Kii Peninsula ALS
Upper motor neuron signs	97%	90%	>95%	80%	>95%
Lower motor neuron signs	100%	40%	100%	60%	100%
Extrapyramidal signs	5%	100%	5%	100%	Rare
Dementia	5%	100%	5%	95%	Rare

Table 14.2. *Comparison of pathological features of high incidence ALS and PD in the Western Pacific*

Neuropathology	Guam ALS	Guam PD	West New Guinea ALS	West New Guinea PD	Kii Peninsula ALS
NFTs	>95%	100%	No autopsies		Common
Senile plaques	Rare	Rare	–		Rare
Lewy Bodies	3–4%	10%	–		–
Neuronal loss					
Brain	Mild	Severe	–		Rare
Cord	Severe	Rare	–		Severe

PD, like ALS, demonstrates the accumulation of abundant Alzheimer NFTs in the brain that are a hallmark neuropathological change characteristic of these disorders (Table 14.2) (Hirano *et al.*, 1961*b*; Garruto, 1991).

Early environment: migration studies

Between 1898 and 1950, strict military control of the island, geographical isolation, and lack of civilian transportation limited the opportunity of Guamanians to travel. With congressional passage of the Organic Act in 1950, Guamanians became citizens of the United States, and a civilian government was established. This led to dramatic out-migration of Chamorros to the United States mainland to vast immigration of non-Chamorros to Guam.

Epidemiological surveillance of Chamorro migrants established that ALS developed after long-term absences from Guam (Garruto, Gajdusek & Chen, 1980). The minimum period of residence on Guam prior to onset of disease was from birth to 18 years of age. The longest period of absence from Guam before onset of disease was over three decades. Based on

Table 14.3. *Percentage prevalence of upper and lower motor neuron involvement in 124 PD patients*

Clinical	Number of patients	% involvement
PD Only	23	18.5%
PD+UMD	44	35.5%
PD+LMD	13	10.5%
PD+UMD+LMD	44	35.5%

From K.M. Chen, 1979.

post-World War II estimates of the number of Chamorro migrants who left Guam, the crude average annual mortality rate for ALS was 5–7 per 100 000 population. The mortality rates were significantly less than the rates for non-migrant Chamorros living on Guam. Thus, Chamorro migrants appeared less likely to develop disease than did those who remained on Guam, but their risk of developing ALS was still 5–10 times greater than that for the general population of the continental United States.

Although Filipinos and other ethnic groups migrated to Guam during the eighteenth century in an effort by the Spanish to bolster the decimated Chamorro population, it was not until after World War II that a resurgence in migrants (particularly of Filipino males) to Guam occurred. Epidemiological studies of ALS and PD in Filipino migrants to Guam established that they migrated from the Ilocos region of northwestern Luzon (Garruto, Gajdusek & Chen, 1981). All Filipino migrants who developed disease were men who had married Chamorro women and lived a traditional lifestyle in Chamorro villages. Their mean number of years of residence on Guam prior to the onset of ALS was 17, and their mean age at migration was 38 years. PD has also developed in these migrants, but as in Chamorro migrants, the number of cases is small (Garruto *et al.*, 1981). Based on the surveillance data available, the average annual crude mortality rate for Filipino migrants is 8 per 100 000 population, a rate similar to that for Chamorro migrants. A survey of ALS and PD in the Ilocos region of the Philippine Islands did not disclose any unusual prevalence rates in the parent population (R. Yanagihara and R.M. Garruto, unpublished observations). Thus the increased risk to Filipino migrants was likely acquired during their long-term residence on Guam. Short-term transient American military personnel and US construction workers and their dependents did not show any increased prevalence of ALS or PD while stationed on the island (Brody, Edgar & Gillespie, 1978).

The importance of these migration studies is that it establishes a minimum latency period for exposure to environmental risk factors and the development of neuronal degeneration and neurofibrillary tangle formation. The increased disease frequency in Chamorro migrants from Guam, many of whom never returned to the island prior to onset of their disease, argues for the involvement of early life events, while disease occurrence in Filipino migrants to Guam argues for the involvement of later life events. Both studies clearly support the concept of a long-temporal sequence, whether initiated in early life or during the early to mid-adult years. In both cases the temporal sequence is approximately two decades.

Early environment: disappearance of high incidence

Since the initiation of systematic studies of ALS and PD on Guam in the 1950s, a number of epidemiological changes have occurred. When ALS on Guam was first described in the mid-1950s, the incidence rate of the disease was reported to be 50-fold higher than the rate of 1/100 000 population for the continental United States with 1 in 5 people over age 25 dying from these disorders (Brody & Kurland, 1973). Today, the risk of developing either disease on Guam has declined dramatically to levels only several-fold higher than that for non-Chamorro residents living in the continental United States (Figure 14.2) (Garruto, Yanagihara & Gajdusek, 1985; Plato et al., 1986).

In addition, the age at onset for both ALS and PD in males and females has increased significantly during this dramatic decline in incidence, while the duration of ALS has decreased and that for PD increased (Table 14.4). These changes are now much more consistent with sporadic ALS worldwide and represent, in part, a secular increase in the longevity of the Chamorro population. In Table 14.5, the age at onset and duration of ALS and PD in West New Guinea and the Kii Peninsula foci are shown for comparison.

In addition to decreases in overall incidence of these disorders, there have also been geographical differences in the decline in the incidence of ALS and PD on Guam (Figure 14.3). The sharpest decline in incidence occurred in the southern villages where the disorders were most prevalent. In the moderate incidence regions of central and northern Guam ALS and PD have been slower to decline, while the western part of the island had a low incidence of both disorders throughout the 35-year surveillance period (Garruto et al., 1985).

The male:female ratio of ALS and PD has also changed. In the 1950s the male:female ratio was 2:1 for ALS and almost 3:1 for PD. Presently, the sex ratio for both ALS and PD approaches unity, with the most significant changes occurring in males (Garruto et al., 1985).

Table 14.4. *Changing patterns in age at onset and duration of Guam ALS and PD*

Diagnosis	Age at onset (years)			Duration (years)	
	1950	1980	Change	1950	1980
Guam ALS					
Male	47.6	51.9	4.3	5.5	3.4
Female	42.1	52.5	10.4	8.0	3.9
Guam PD					
Male	55.5	59.5	4.0	4.5	5.6
Female	50.9	58.9	8.0	4.0	5.9

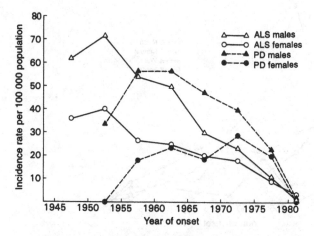

Figure 14.2. Disappearance of amyotrophic lateral sclerosis (ALS) and parkinsonism-dementia (PD) of Guam. Five year average annual incidence rates in Chamorro males and females by year of onset. Rates are age and sex adjusted to the 1960 Guamanian Chamorro population.

Equally important has been the declining incidence and prevalence rates of ALS and PD in the remaining two Pacific foci (Table 14.6). In the Kii peninsula focus of Japan, incidence rates in the Hobara subfocus have decreased from 55 to 14 per 100 000 and no new cases have been seen in the Kozagawa subfocus during the past ten years (Garruto & Yase, 1986; Shiraki & Yase, 1991). In the remote West New Guinea focus, with village prevalence rates as high as 1300 per 100 000 population, ALS and PD have also disappeared during the past two decades in one village and declined in several others that have undergone increasing acculturation with the associated introduction of new foodstuffs (Gajdusek, 1984; Gajdusek & Salazar, 1982; D.C. Gajdusek, personal communication). These dramatic

Table 14.5. *Age at onset and duration of ALS and PD in West New Guinea and the Kii Peninsula of Japan*

Region/diagnosis	Age at onset (years)	Duration (years)
West New Guinea		
ALS	33	4
PD	43	6
Kii Peninsula		
ALS	50	4

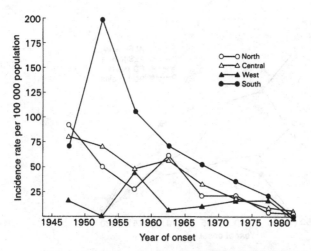

Figure 14.3. Decline by geographic region of amyotrophic lateral sclerosis (ALS) in Chamorro males represented by five-year average annual incidence rates. Similar declines in the same regions occurred in ALS females and in parkinsonism-dementia males and females.

epidemiological changes strongly support the possibility of a cohort effect and a geographical distribution of disease based on place of residence. The involvement of environmental factors contributing to the expression of these diseases has become a virtual certainty in view of the striking decline in the incidence rates. It also suggests that, if genetics are involved, they are likely to be secondary to some environmental insult, although a recent study by Bailey-Wilson and colleagues (1993) cannot exclude a major gene effect in combination with an environmental trigger.

A unified mechanism proposed by Garruto (1991), states that a basic defect in mineral metabolism induces a form of secondary hyper-parathyroidism, provoked by chronic nutritional deficiencies of calcium,

Table 14.6. *Comparison of incidence/prevalence rates among high incidence Pacific Foci of ALS and PD*

Height of epidemic* (Rate per 100 000 population)		Today (Rate per 100 000 population)
Guam		
ALS	55	≈ 5
PD	40	≈ 5
West New Guinea		
ALS	147	Dramatic decline in villages with western contact
PD	79	(no rates available)
Kii Peninsula		
ALS	55	14

*Highest incidence villages on Guam ≈ 350/100 000 population while in West New Guinea some villages as high as 1300/100 000 population.

leading to enhanced gastrointestinal absorption of bioavailable aluminium and the deposition of calcium and aluminium in neurofibrillary-tangle (NFT)-bearing neurons, the hallmark neuropathological lesion in these disorders (Garruto, 1991). Through alterations in normal biosynthesis, phosphorylation and/or catabolism of cytoskeletal proteins and brain amyloid, this elemental deposition could lead to a fundamental disruption in slow axonal transport. Empirical support for this unified mechanism is summarized by Garruto (1991) and includes the finding of low calcium and high aluminium levels in the environment of all three foci and the dramatic intraneuronal elemental deposition of calcium and aluminium in brain and spinal cord tissues of these patients (Figure 14.4). Increased intake of calcium in later life and the subsequent correction of any secondary hyperparathyroidism is unlikely to prevent or reverse the neuronal damage and cell death that these deposits are likely to have produced.

The changing disease patterns are consistent with a population undergoing significant cultural, nutritional and economic transitions. The disappearance of high incidence ALS and PD on Guam has occurred during the aggressive acculturation from a previously horticultural and fisherfolk subsistence economy immediately following World War II, to a westernized culture with a cash economy. Economic developments occurred faster in the northern and central regions of the island, where large military installations, major harbour facilities and the capital city of Agana are located. The high-incidence southern region was more impervious to change, but increasingly, men living in the south went to work in the more economically developed northern and central regions. Women in this

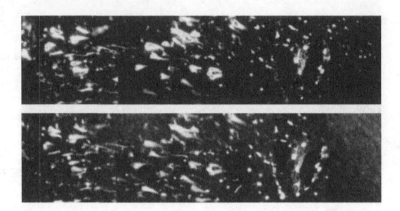

Figure 14.4. Composite of elemental images using computer controlled X-ray microanalysis and wavelength dispersive spectrometry of the pyramidal cell layer of the hippocampus in a patient with parkinsonism-dementia. The images show the striking co-localization of calcium (top) and aluminium (bottom) in the same neurofibrillary tangle-bearing neurons. The field size is 1.0 mm × 1.0 mm for each of the six individual images in the composite for each element. For details see Garruto et al., 1984.

oceanic Latin culture were much slower to follow this trend and continued to remain at home. This may account for the earlier and sharper decline in disease incidence among men, particularly those from traditional southern villages (Figure 14.3).

Finally, in more than three decades of searching, a satisfactory genetic explanation for ALS and PD has not been found (Garruto, 1991), including a recent search for point mutations in the exon regions of the SOD1 (Figlewicz et al., 1994) and SOD2 genes (D. Figlewicz, unpublished observations). No other environmental agent, including conventional and unconventional viruses and other toxins such as that proposed for the nut of the false sago palm (*Cycas circinalis*), has been found to be causally associated with the disease process (Gibbs & Gajdusek, 1982; Duncan et al., 1988, 1990; Garruto, Yanagihara & Gajdusek, 1988; Garruto et al., 1989; Gajdusek, 1990).

Neurofibrillary tangle formation: a neuropathological link and key to understanding the disease process

As mentioned earlier, the hallmark neuropathological lesions in Guamanian ALS and PD are NFTs of the brain and spinal cord (Garruto, 1991; Hirano, 1961b). These lesions form a neuropathological link between ALS, PD, late Down's Syndrome and Alzheimer's disease, but unlike the latter two diseases, are found without associated senile plaques. It is also rather astounding that, during the height of the 'epidemic' of ALS and PD, nearly 70% of the clinically normal population of Guam over age 25 developed NFTs, indicating a preclinical disease state (Anderson *et al.*, 1979).

At the light microscopic level, NFTs are argentophilic and congophilic–birefringent under polarized light. They are intraneuronal and composed of cytoskeletal and other proteins which aggregate into fibrillary bundles giving a tangle-like appearance (Figure 14.5). They are abundant in number and appear in a specific, yet widespread, distribution throughout the brain, and in Guamanian ALS, in the spinal cord as well (Hirano, 1973). NFTs in Guamanian ALS and PD are indistinguishable from those seen in Alzheimer's disease, which are usually flame-shaped in the cortex and more globular in appearance in the brainstem. Once an NFT-bearing neuron dies, the fibrillary material remains as a permanent marker or 'tombstone' indicating its former presence.

Major fibrillary components of the NFTs in Alzheimer's disease are paired helical filaments. By electron microscopy, these filaments are 8–12 nm in diameter and helically wound with a crossover periodicity of 78–80 nm (Figure 14.6) (Kidd, 1963; Hirano, 1973; Wisniewski, Narang & Terry, 1976; Fukatsu *et al.*, 1988). In Guamanian PD, purified preparations of NFTs consist mainly of paired helical filaments similar to those found in Alzheimer disease, while in ALS single and occasionally triple straight filaments predominate (Guiroy *et al.*, 1987, 1993). Ultrastructurally, the paired helical filaments from Guamanian PD patients measure 10 nm in diameter and are helically wound with an average crossover periodicity of 180 nm (Figure 14.7), more than twice the crossover periodicity of paired helical filaments in Alzheimer's disease. The significance of this more loosely wound helical twist in NFTs in PD is not known. Each of the single filaments in PD appear to be composed of two or more subfilaments 2–5 nm in diameter, similar to those in Alzheimer's disease, while in ALS the single filaments are 5–20 nm in diameter and at higher magnification show extensive braiding of 3–4 subfilaments.

Immunocytochemically, several cytoskeletal proteins, including neurofilament, microtubule-associated protein-2 and -tau and ubiquitin and amyloid β-protein have been found in the NFTs of brain and spinal cord tissues of Guamanian patients with ALS and PD as well as in neurologically normal Guamanians with early neurofibrillary degeneration (Shankar *et al.*, 1989; Matsumoto, Hirano & Goto, 1990*a*,*b*; Guiroy *et al.*, 1993; S.K. Shankar, unpublished data). The immunocytochemical pattern of NFTs in Guamanian ALS and PD closely resembles that seen in Alzheimer's disease.

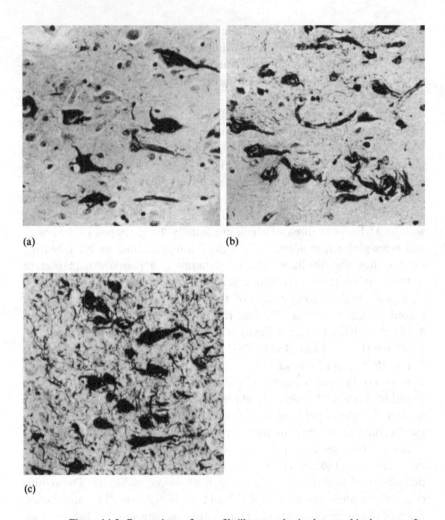

(a)   (b)

(c)

Figure 14.5. Comparison of neurofibrillary tangles in the entorhinal cortex of patients with (a) amyotrophic lateral sclerosis (b) parkinsonism-dementia and (c) Alzheimer's disease. Gallyas silver stain. × 350. From Wakayama *et al.*, 1993.

Biochemical and molecular characterization of purified preparations of NFTs from Guamanian patients with ALS and PD and in neurologically normal Guamanians with NFTs has clearly demonstrated that these lesions contain multimeric aggregates of a low molecular weight (4.0–4.5 kD) 42 or 43 amino acid hydrophobic polypeptide that forms para-crystalline arrays of β-pleated sheets of amyloid β- protein (Guiroy *et al.*, 1987, 1993). In PD the amino acid composition of this formic acid-soluble

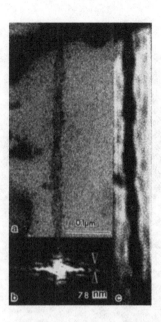

Figure 14.6. Electron micrograph of paired helical filaments in Alzheimer's disease (a) from thin sections, (b) enhanced by computer digital processing showing cross-over periodicity and (c) noise filtered image of (a) showing rotational symmetry. Preparation stained with uranyl acetate. × 200 000. From Fukatsu *et al.*, 1988.

protein is similar to that of Alzheimer's disease, although only 60% are full-length amino acid chains (Guiroy *et al.*, 1987). However, the *N*-terminal amino acid sequence of this polypeptide is identical to that of the brain amyloid in Alzheimer's disease and late Down's syndrome.

It seems certain that the key to understanding Pacific ALS and PD is through the understanding of the biosynthesis, phosphorylation and/or catabolism of these proteins. The NFTs are complex morphological, immunocytochemical and molecular structures. They are composed of several proteins that likely result from aberrant post-translational processing, phosphorylation and co-polymerization of neurofilament, tau and amyloid β-protein that leads to neuronal impairment, neuronal degeneration and eventual cell death.

Protein fibrillization: a long-term process of protein polymerization and molecular reorganization

The process of protein fibrillization leading to the formation of NFTs involves a long temporal sequence that may begin *in utero*, during infancy,

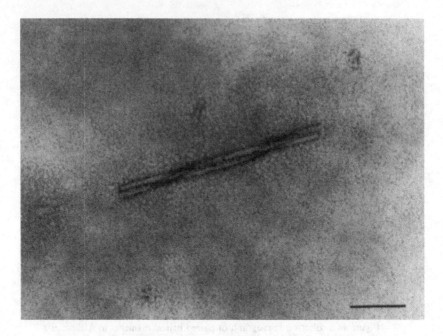

Figure 14.7. Electron micrograph of paired helical filaments from purified neurofibrillary tangles in a Guamanian patient with parkinsonism-dementia. Preparation stained with uranyl acetate. Bar = 100 nm. From Guiroy et al., 1987.

childhood or adolescence or even later, perhaps spanning as much as several decades prior to onset of clinical disease. Regardless of whether the initiating event is a point mutation in the amyloid precursor protein; a neurotoxin such as aluminium that interferes with biosynthesis, phosphorylation of catabolism of cytoskeletal proteins (neurofilament and microtubule associated protein tau); a nucleating event as a mechanism for protein self-replication and polymerization; or some other initiating event, the process of long-term fibrillization can be conceived of as the slow accumulation of protein molecules that go undetected in preclinical states moving from solubilized to insoluble states forming single fibrils that develop rearranged configurations often into β- pleated sheets. Thus, the timing of the onset of 'non-infectious' brain amyloid forming disorders may be predictable, based on the speed with which each initiating event leads to fibrillization (Figure 14.8). In Down's syndrome, this is represented by an over-production of amyloid precursor protein due to a triplet rather than a doublet gene for the precursor protein. In addition, onset of Alzheimer's disease pathology in patients with late Down's syndrome occurs at an early age (25–35 years), since the process of molecular accumulation of the amyloid β-protein is likely to start *in utero* or at birth. Where the initiating sequence is induced by a toxin such as aluminium, the process could occur later in life and with chronic exposure to low toxic doses which act slowly, possibly resulting in a longer temporal sequence with clinical onset in mid-life (45–55 years). If the process is initiated by point mutations on chromosomes 14, 19 or 21 as in familial Alzheimer's disease, this may represent an early life event of molecular accumulation of amyloid β-protein with a resultant clinical onset in mid-life. At the opposite end of the temporal scale, sporadic Alzheimer's disease in late life (70–80 years) may result from a much slower process of protein fibrillization (including microtubule associated protein tau) with an uncertain timing in its preclinical initiation (Figure 14.8).

The theoretical construct shown in Figure 14.8 is just one way to imagine the development of an entire class of NFT-forming disorders with long-latency and slow progression. This same sort of construct also can be applied to what is known about Chamorro migrants from Guam and Filipino migrants to Guam, who develop disease within a two to three-decade timeframe after exposure to the Guam environment, with Chamorro migrants exposed early in life and Filipino migrants exposed later in life. Nearly 70% of the clinically normal adult Chamorro population of Guam develop NFTs and, of these, 20% go on to develop ALS and PD. This is likely to represent an extraordinary genetic susceptibility to the development of NFTs, or to be the result of some dose-dependent exposure to an environmental agent. In any case, if this process represents single or multiple insults that can be delayed, turned off, or eliminated prior to reaching a critical threshold during this long temporal sequence, then no clinical expression of disease would be likely to occur (Figure 14.9).

Human and experimental evidence of aluminium intoxication and neuro-degeneration

Aluminium as a systemic toxin and as a neurotoxin has been implicated in human disorders other than Pacific ALS and PD for many years. In Alzheimer's disease, Alexander and colleagues (1937), using a method of microincineration, appear to be the first to demonstrate that Alzheimer NFTs contain 'ample mineral deposits in the neurofibrillary strands' which were 'white and for the most part non-soluble in water'. This and other early studies set the stage for the report by Crapper, Krishnan & Dalton

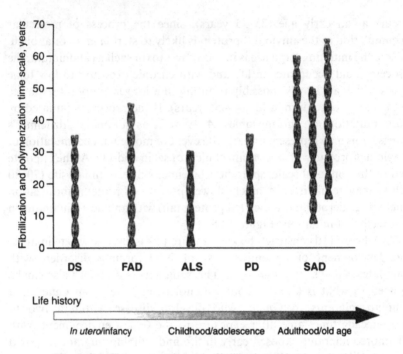

Figure 14.8. The process of protein fibrillization and polymerization in neurodegenerative disorders of long-latency and slow progression. The initial insult can be environmental or genetic and occur anytime during the life cycle. The number of years necessary for the process of fibrillization to occur sufficiently enough to produce clinical disease in measured in decades and can be accelerated or decelerated by the magnitude of the insult, the chronicity of the insult and the age at which the insult begins. If the process is turned off or delayed or the insult removed, clinical expression of disease may not occur (See Figure 14.9). DS = Down's syndrome; FAD = familial Alzheimer's disease; ALS = amyotrophic lateral sclerosis; PD = parkinsonism-dementia; SAD = sporadic Alzheimer's disease.

(1973) of increased concentration of aluminium in the brains of patients with Alzheimer's disease. A year earlier, Nikaido and colleagues (1972) presented the first evidence of silicon in Alzheimer senile plaques using X-ray microanalysis. This was followed by demonstration of intracellular deposition of aluminium in NFT-bearing neurons (Perl & Brody, 1980) and in senile (amyloid) plaque cores (Candy *et al.*, 1986; Masters *et al.*, 1985). However, the implication of aluminium as a risk factor in Alzheimer's disease remains highly controversial.

By contrast, the evidence for aluminium intoxication in dialysis encephalopathy is overwhelming (Alfrey, Legendre & Kaehny, 1976; Bakir *et al.*, 1986; Molitoris *et al.*, 1989). Initially, the source of the aluminium responsible for the toxicity was primarily the water used to prepare the

Single/multiple switch (insult) theory

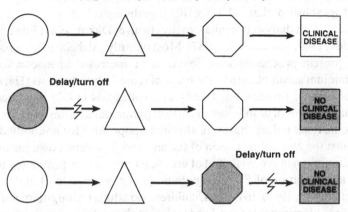

Figure 14.9. Diagrammatic representation of the long latency and slow progression of amyotrophic lateral sclerosis (ALS) and parkinsonism-dementia (PD), exemplified by the development of neurofibrillary tangles where 70% of the Chamorro population develop lesions while only 20% go on to develop clinical disease. The top line of the Figure demonstrates an uninterrupted process leading to clinical disease. The second and third lines of the Figure represent an interruption in the disease process that results in no clinical disease. The latter lines may represent what happened during the disappearance of the high incidence of ALS and PD in all three Pacific foci; this is characterized by a single or multiple insult event, for example, a genetic susceptibility (metabolic defect) and an environmental trigger (aluminum). If one of these insults is delayed or turned off, clinical onset of ALS or PD does not occur.

dialysate. Subsequently, however, a second source of aluminium was found to contribute to the toxicity, namely the orally administered aluminium hydroxide used as a phosphate binding agent. Most recently, citrate and other low molecular weight organic ligands have been shown to markedly enhance the gastrointestinal absorption of orally administered aluminium compounds (Froment *et al.*, 1989; Molitoris *et al.*, 1989; Slanina *et al.*, 1986; Weberg & Berstad, 1986). Analogous to Guamanian ALS and PD patients, in whom a metabolic or barrier defect and secondary hyper-parathyroidism have been postulated as mechanisms of increased absorption of bioavailable aluminium, uraemic patients are subject to aluminium loading because of compromised renal function, resulting in osteomalacia secondary to aluminium toxicity.

Earlier, many have argued that the established causal role of aluminium in dialysis encephalopathy was not relevant to a discussion of aluminium's role in neuro-degenerative disorders such as Alzheimer's disease, mainly because of the failure to find Alzheimer lesions in dialysis encephalopathy, namely NFTs, senile plaques and amyloid β-protein deposition. It is now well established that all three neuropathological changes do occur in patients with dialysis encephalopathy (Brun & Dictor, 1981; Edwardson & Candy, 1989; Scholtz et al., 1987). Most recently, Alzheimer-like changes in tau protein processing were found to be increased in association with aluminium accumulation in the brain of renal dialysis patients (Harrington et al., 1994). In addition, Candy and colleagues (1992) identified focal accumulations of aluminium in cortical pyramidal neurons in patients with chronic renal failure that were also immunopositive for antibody directed against the N-terminal region of the amyloid β-protein. Also, the presence of aluminium in synovial fluid of uremic patients is now postulated to cause the accumulation of β_2-microglobulin (a systemic amyloid) in their joints (Netter et al., 1990). In infants, children and adults on long-term parenteral feeding, aluminium is found to localize at the surface of mineralized bone (Berquist, Ament & Coburn, 1982; Klein et al., 1982a,b; Ott et al., 1983; Sedman et al., 1985). Similarly, premature infants receiving chronic intravenous therapy develop progressive osteopenia and fractures with aluminium loading of bone from parenteral solutions (Klein et al., 1982a,b; Sedman et al., 1985; Gruskin, 1991; Sedman, 1992). In addition, infants with congenital uraemia fed commercial infant formula contaminated with aluminium may develop aluminium toxicity with high levels of aluminium content in their brain tissues (Freundlich et al., 1985). These studies have been recently augmented by a report demonstrating increased aluminium content in cortical grey matter of a preterm infant receiving parenteral feeding solutions contaminated with aluminium (Bishop et al., 1989).

Preterm infants are at increased risk of aluminium loading because of reduced renal function and high aluminium concentrations in feeding solutions. Also, healthy preterm infants (< 33 weeks) fed orally have been reported to have about twice the serum aluminium levels as full-term infants, despite equal aluminium intakes (Bougle et al., 1992). The most likely explanation is poorer renal clearance in preterm infants, although increased aluminium absorption during the first weeks of life also has been suggested (Sedman et al., 1985), but this has been debated (Owen, 1989). Finally, quite dramatic increases (about ninefold) in serum aluminium have been reported in normal infants of mean age 6 months who had received antacid therapy for symptomatic relief of gastroesophageal reflux (Tsou et al., 1991).

Thus, there appears to be sufficient evidence during early life to suggest an increased risk of developing disease in later life. One of the mitigating factors as explained in the previous section are the conditions under which this process proceeds. In the case of significantly barrier breeches such as kidney and gut, or underdeveloped organ function such as kidney function of preterm infants, the process of neuro-degeneration can be greatly accelerated, resulting in a much shorter latency of weeks, months, or years, rather than decades prior to onset of clinical disease.

In aged adult rabbits, aluminium is more toxic and there is a much higher mortality than in younger rabbits using repeated injections of aluminium (Wisniewski, Sturman & Shenk, 1980; Petit *et al.*, 1985; Yokel, 1989). Adult rabbits also demonstrate a more pronounced aluminium-induced behavioural toxicity compared to young rabbits, despite similar tissue aluminium concentrations. However, young rabbits are more susceptible to aluminium-induced skeletal toxicity than adult rabbits (Yokel, 1987) and fetuses are more susceptible to transplacental toxicity through inoculation of their mothers than neonates receiving mother's milk (Yokel, 1984, 1985).

Aluminium chloride administered orally to pregnant rats has been shown to result in delayed learning acquisition in the offspring (Misawa & Shigeta, 1993). Other reports also show that maternal dietary exposure during gestation and/or lactation can result in developmental alterations in young rats and young mice (Bernuzzi, Desor & Lehr, 1986; 1989; Golub *et al.*, 1987; Golub, Keen & Gershwin, 1992; Donald *et al.*, 1989; Muller *et al.*, 1990; Clayton *et al.*, 1992). Golub and colleagues (1992) reported reduced forelimb grasp strength and negative geotaxis in offspring of mice, particularly after gestational aluminium exposure. Donald & colleagues (1989) noted that maternal dietary exposure to excess aluminium during gestation and lactation results in persistent neurobehavioural deficits in weanling mice, even when maternal toxicity is not present. Rat pups treated by gastric intubation with aluminium lactate show transitory delay of neuromotor development (Bernuzzi *et al.*, 1989) and reduction in choline acetyltransferase activity (Cherroret *et al.*, 1992). Intraperitoneal injections of aluminium sulfate to pregnant mice lead to retarded neurobehavioural development in the offspring, as well as a reduction in the 'rate of ultrasonic calling', a suggested sensitive measure of behavioural teratogenicity (Rankin & Manning, 1993). Finally, Gomez and colleagues (1991) reported that aluminium citrate, but not aluminium hydroxide, given to pregnant rats increased the incidence of skeletal malformations in their offspring.

As a direct outgrowth of long-term systematic studies of ALS and PD in the Western Pacific, experimental models of chronic aluminium-induced neuronal degeneration and NFT formation in non-human primates and rabbits and models of neuronal degeneration and neurofibrillary pathology in primary neuronal cell cultures have been developed (Garruto, 1991). In non-human primates it has been demonstrated that low-dose aluminium salts administered orally to juvenile macaques maintained on a low calcium diet for 12–46 months, while not inducing clinical disease, resulted in a selective motor neuron pathology with localization of aluminium in the spinal cord at autopsy (Garruto *et al.*, 1989; Yasui *et al.*, 1991; Yano *et al.*, 1989). It is suspected that, had these primates with normal kidney and gastric function remained on test longer (perhaps several more years), they may well have developed overt clinical disease. Other studies using rabbits (Strong *et al.*, 1991; Kihara *et al.*, 1994) and primary neuronal cultures (Strong & Garruto, 1991) support the concept of long-term selective nervous system degeneration using long-term low dose administration of aluminium salts.

The above human and experimental findings have led to a search for new models in nature that would complement and expand these observations. Aluminium intoxication is not a new phenomenon and has been a recognized problem in plants for decades (Foy, Chaney & White, 1978; Hutchinson, 1943; Hartwell & Pember, 1918; Roy, Sharma & Talukder, 1988). Soluble aluminium from soil is known to impair the growth and development of many plant species, including domestic crops such as coffee and wheat (Parker, Zelazny & Kinraide, 1989; Pavan, Bingham & Pratt, 1982). Likewise, in certain regions of the world, forests have been in dramatic decline as a direct result of soluble aluminium. Among the trees most heavily affected are evergreens, particularly spruce and fir, because bioavailable aluminium inhibits calcium uptake and causes a decline in cambial growth (Shortle & Smith, 1988). This process is particularly damaging to older trees that have an increased demand for calcium; they lose sapwood and become increasingly vulnerable to insects and various pathogenic microorganisms. The interaction between calcium and aluminium continues to be a repetitive theme in many studies.

In the animal world, it is now well established that aluminium is the main toxicant leading to fish extinction in areas where acid rain has leached the metal from minerals in the watershed soil (Driscoll *et al.*, 1980; Schofield & Trojnar, 1980). Such watersheds are vulnerable mainly because of lack of calcium minerals and thus provide no buffering capacity to neutralize the acid precipitation. The main target organ of aluminium in mature fish are the gills. Previously, the nervous system of these fish had never been examined for aluminium-induced neuropathological changes. In acid lakes, fish gills and primary receptor neurons in the olfactory neuropathway are in direct and continuous contact with water containing up to several hundred micrograms of soluble aluminium per litre, levels that, when found in the blood of humans with chronic renal failure, rapidly cause dialysis encephalopathy.

As a result, our research efforts led to the study of natural models of chronic aluminium intoxication of the central nervous system in fish from acid rain lakes in the United States. In the Adirondack Park in northern New York State, various species of fish from lakes with different aluminium concentrations and water chemistries were studied (Flaten *et al.*, 1993). The fish were processed at the field sites by immediate dissection and quick freezing of tissues on dry ice, or by whole-body perfusion. The central nervous system of fish from aluminium-rich lakes had neuro-pathological changes similar to those found in experimental aluminium intoxication. Deposition of aluminium was found not only in the primary lamellae of gills but in the olfactory epithelium and in the olfactory bulb (Flaten *et al.*, 1993). This is especially interesting in light of the involvement of the olfactory system in neuro-degenerative disorders such as Guamanian ALS and PD, Parkinson's disease and Alzheimer's disease (Doty, Deems & Stellar, 1988; Doty *et al.*, 1991; Elian, 1991). These studies, still in their infancy, hold significant promise for the understanding of the cellular and molecular mechanisms of aluminium intoxication under natural and experimental conditions.

Conclusions

In neuro-degenerative disorders with late onset and slow progression, it is almost certain that the process of degeneration has a long temporal sequence initiated not only in early life, but throughout the lifespan. The three most important factors influencing the onset of neuro-degenerative disorders may be age, time and the sequence of events. A removal or delay in the initiating insults or events may delay or prevent the clinical expression of disease. The exact process by which this takes place is uncertain and specific to a particular neuro-degenerative disorder. However, there is a limited (rather than an infinite) number of responses by the nervous system to various types of insults, irrespective of whether such insults are genetic or environmental, which leads to concepts of disease continuums and disease syndromes with similar pathogenetic courses.

The examples of neuro-degenerative diseases set forth in this chapter represent an interesting interrelationship at the pathological, cellular and molecular level. To what extent Pacific ALS and PD and Alzheimer's disease are related disease entities is not yet completely clear, but the process of fibrillization in each of these disorders represents a long temporal sequence that is measured in decades rather than weeks, months and years and results in similar lesions (NFTs) that lead to neuronal dysfunction and ultimately cell death. When the cumulative effects of this long temporal process are fully realized, clinical expression as an end stage disease process is finally observed. The extent to which the scientific community can successfully identify and record life history events, and gauge their role in adult onset disorders is at the moment difficult to evaluate, but the question of whether to put forth the effort to do so can be clearly answered.

Acknowledgements

I would like to thank Linda Poole and Virginia Rousculp for their editorial assistance and Dr Trond Peder Flaten for reviewing the manuscript and making critical suggestions.

References

Alexander, L., Myerson, A. & Goldman, D.E. (1937). The mineral content of various cerebral lesions as demonstrated by the microincineration method. *American Journal of Pathology*, **13**, 405–39.
Alfrey, A.C., Legendre, G.R. & Kaehny, W.D. (1976). The dialysis encephalopathy syndrome: possible aluminium intoxication. *New England Journal of Medicine*, **294**, 184–8.
Anderson, F.H., Richardson, E.P., Jr., Okazaki, H. & Brody, J.A. (1979). Neurofibrillary degeneration on Guam: frequency in Chamorros and non-Chamorros with no known neurological disease. *Brain*, **102**, 65–77.
Arnold, A., Edgren, D.C. & Palladino, V.S. (1953). Amyotrophic lateral sclerosis: fifty cases observed on Guam. *Journal of Nervous Mental Disorders*, **177**, 135–9.
Bailey-Wilson, J.E., Plato, C.C., Elston, R.C. & Garruto, R.M. (1993). Potential role of an additive genetic component in the etiology of amyotrophic lateral sclerosis and parkinsonism-dementia in the western Pacific. *American Journal of Medical Genetics*, **45**, 68–76.
Bakir, A.A., Hryhorczuk, D.O., Berman, E. & Dunes, G. (1986). Acute fatal hyperaluminemic encephalopathy in undialyzed and recently dialyzed uremic patients. *Transactions of the American Society of Artificial Internal Organs*, **32**, 171–6.
Bernuzzi, V., Desor, D. & Lehr, P.R. (1986). Effects of prenatal aluminium exposure on neuromotor maturation in the rat. *Neurobehavioral Toxicology and Teratology*, **8**, 115–19.
Bernuzzi, V., Desor, D. & Lehr, P.R. (1989). Developmental alterations in offspring of female rats orally intoxicated by aluminum chloride or lactate during gestation. *Teratology*, **40**, 21–7.
Berquist, W.E., Ament, M.E. & Coburn, J.W. (1982). Aluminum as a factor in the bone disease of long-term parenteral nutrition. *Transactions of the Association of American Physicians*, **95**, 155–64.
Bishop, N.J., Robinson, M.J., Lendon, M., Hewitt, C.D., Day, J.P. & O'Hara, M. (1989). Increased concentration of aluminum in the brain of a parenterally fed preterm infant. *Archives of Disease in Childhood*, **64**, 1316–17.
Bougle, D., Bureau, F., Voirin, J., Neuville, D. & Duhamel, J.F. (1992). A cross-sectional study of plasma and urinary aluminum levels in term and preterm infants. *Journal of Parenteral and Enteral Nutrition*, **16**, 157–9.

Brant, L.J. & Pearson, J.D. (1994). Modeling the variability in longitudinal patterns of ageing. In *Biological Anthropology and Aging: Perspectives on Human Variation Over the Life Span*, ed. D.E. Crews and R.M. Garruto, pp. 373–93. New York: Oxford University Press.

Brody, J.A., Edgar, A.H. & Gillespie, M.M. (1978). Amyotrophic lateral sclerosis. No increase among US construction workers in Guam. *Journal of the American Medical Association*, **240**, 551–2.

Brody, J.A. & Kurland, L.T. (1973). Amyotrophic lateral sclerosis and parkinsonism-dementia in Guam. In *Tropical Neurology*, ed. J.D. Spillane, pp. 355–75.

Brun, A. & Dictor, M. (1981). Senile plaques and tangles in dialysis dementia. *Acta Pathologica et Microbiologica Scandinavica*, **89**, 193–8.

Candy, J.M., McArthur, F.K., Oakley, A.E., Taylor, G.A., Chen, C.P.L.-H., Mountfort, S.A., Thompson, J.E., Chalker, P.R., Bishop, H.E., Beyreuther, K., Perry, G., Ward, M.K., Martyn, C.N. & Edwardson, J.A. (1992). Aluminum accumulation and senile plaque formation in the brains of patients with renal failure, *Journal of the Neurological Sciences*, **107**, 210–18.

Candy, J.M., Oakley, A.E., Klinowski, J., Carpenter, T.A., Perry, R.H., Atack, J.R., Perry, E.K., Blessed, G., Fairbairn, A. & Edwardson, J.A. (1986). Aluminosilicates and senile plaque formation in Alzheimer's disease. *Lancet*, **i**, 354–7.

Chen, K-M. (1979). Motor neuron involvement in parkinsonian dementia and its relationship to Guam ALS. In *Amyotrophic Lateral Sclerosis. Proceedings of the International Symposium on Amyotrophic Lateral Sclerosis*, ed. Japan Medical Research Foundation, pp. 319–44. Tokyo: University of Tokyo Press.

Cherroret, G., Bernuzzi, V., Desor, D., Hutin, M-F., Bernel, D. & Lehr, P.R. (1992). Effects of postnatal aluminum exposure on choline acetyltransferase activity and learning abilities in the rat. *Neurotoxicology and Teratology*, **14**, 259–64.

Clayton, R.M., Sedowofia, S.K.A., Rankin, J.M. & Manning, A. (1992). Long-term effects of aluminium on the fetal mouse brain. *Life Sciences*, **52**, 1921–8.

Crapper, D.R., Krishnan, S.S. & Dalton, A.J. (1973). Brain aluminum distribution in Alzheimer's disease and experimental neurofibrillary degeneration. *Science*, **180**, 511–13.

Donald, J.M., Golub, M.S., Gershwin, M.E. & Keen, C.L. (1989). Neurobehavioral effects in offspring of mice given excess aluminum in diet during gestation and lactation. *Neurotoxicology and Teratology*, **11**, 345–51.

Doty, R.L., Deems, D.A. & Stellar, S. (1988). Olfactory dysfunction in parkinsonism: a general deficit unrelated to neurologic signs, disease stage, or disease duration. *Neurology*, **38**, 1237–44.

Doty, R.L., Perl, D.P., Steele, J.C., Chen, K.M., Pierce, J.D., Reyes, P. & Kurland, L.T. (1991). Odor identification deficit of the parkinsonism–dementia complex of Guam: equivalence to that of Alzheimer's and idiopathic Parkinson's disease. *Neurology*, **41** (Suppl. 2), 77–81.

Driscoll, C.T., Baker, J.P., Bisogni, J.J. & Schofield, C.L. (1980). Effect of aluminium speciation on fish in dilute acidified waters. *Nature*, **284**, 161–4.

Duncan, M.W., Kopin, I.J., Garruto, R.M., Lavine, L. & Markey, S.P. (1988). 2-amino-3 (methylamino)-propionic acid in cycad-derived foods is an unlikely cause of amyotrophic lateral sclerosis/parkinsonism. *Lancet*, **ii**, 631–2.

Duncan, M.W., Steele, J.C., Kopin, I.J. & Markey, S.P. (1990). 2-amino-3

244   R.M. Garruto

(methylamino)-propanoic acid (BMAA) in cycad flour: an unlikely cause of amyotrophic lateral sclerosis and parkinsonism dementia of Guam. *Neurology*, **40**, 767–72.

Edwardson, J.A. & Candy, J.M. (1989). Aluminium and the pathogenesis of senile plaques in Alzheimer's disease, Down's syndrome and chronic renal dialysis. *Annals of Medicine*, **21**, 95–7.

Eisen, A.A. & Calne, D.B. (1990). Latent neuro-abiotrophies: A clue to amyotrophic lateral sclerosis. In *Amyotrophic Lateral Sclerosis: Concepts in Pathogenesis and Etiology*, ed. A.J. Hudson, pp. 296–316. Toronto: University of Toronto Press.

Elian, M. (1991). Olfactory impairment in motor neuron disease: a pilot study. *Journal of Neurology, Neurosurgery, and Psychiatry*, **54**, 927–8.

Elizan, T.S., Hirano, A., Abrams, B.M., Need, R.L., Van Nuis, C. & Kurland, L.T. (1966). Amyotrophic lateral sclerosis/parkinsonism dementia complex of Guam. Neurological re-evaluation. *Archives of Neurology*, **14**, 356–68.

Figlewicz, D.A., Garruto, R.M., Krizus, A., Yanagihara, R. & Rouleau, G.A. (1994). Absence of mutations in the Cu/Zn superoxide dismutase gene in amyotrophic lateral sclerosis and parkinsonism-dementia of Guam. *Neuroreport*, **5**, 557–60.

Flaten, T.P., Wakayama, I., Sturm, M.J., Kretser, W., Capone, S., Dudones, T., Bath, D.W., Gallagher, J., Moore, W. & Garruto, R.M. (1993). Natural models of aluminum neurotoxicity in fish from lakes influenced by acid precipitation. In *Heavy Metals in the Environment–9th International Conference*, Vol. **2**, ed. R.J. Allan and J.O. Nriagu, pp. 285–288. Edinburgh: CEP Consultants Ltd.

Foy, C.D., Chaney, R.L. & White, M.C. (1978). The physiology of metal toxicity in plants. *Annual Review of Plant Physiology*, **29**, 511–66.

Freundlich, M., Zilleruelo, G., Abitbol, C. & Strauss, J. (1985). Infant formula as a cause of aluminium toxicity in neonatal uraemia. *Lancet*, **ii**, 527–9.

Froment, D.H., Molitoris, B.A., Buddington, B., Miller, N. & Alfrey, A.C. (1989). Site and mechanism of enhanced gastrointestinal absorption of aluminum by citrate. *Kidney International*, **36**, 978–84.

Fukatsu, R., Obara, T., Baba, N., Kanaya, K., Garruto, R., Hayashishita, T., Kurokawa, Y., Ueno, T. & Takahata, N. (1988). Ultrastructural analysis of neurofibrillary tangles of Alzheimer's disease using computerized digital processing. *Acta Neuropathologica*, **75**, 519–22.

Gajdusek, D.C. (1963). Motor-neuron disease in natives of New Guinea. *New England Journal of Medicine*, **268**, 474–6.

Gajdusek, D.C. (1984). Environmental factors provoking physiological changes which induce motor neuron disease and early neuronal ageing in high incidence foci in the western Pacific. In *Research Progress in Motor Neuron Disease*, ed. F.C. Rose, pp. 44–69. Kent: Pitman Books Ltd.

Gajdusek, D.C. (1990). Cycad toxicity not the cause of high incidence amyotrophic lateral sclerosis/parkinsonism dementia on Guam, Kii peninsula of Japan, or in West New Guinea. In *Amyotrophic Lateral Sclerosis: Concepts in Pathogenesis and Etiology*, ed. A.H. Hudson, pp. 317–25. Toronto: University of Toronto Press.

Gajdusek, D.C. (1994). Spontaneous generation of infectious nucleating amyloids in the transmissible and non-transmissible cerebral amyloidoses. *Molecular*

Neurobiology, **8**, 1–13.

Gajdusek, D.C. & Salazar, A.M. (1982). Amyotrophic lateral sclerosis and parkinsonian syndromes in high incidence among the Auyu and Jakai people in West New Guinea. *Neurology*, **32**, 107–26.

Garruto, R.M. (1991). Pacific paradigms of environmentally induced neurological disease: clinical, epidemiological and molecular perspectives. *Neurotoxicology*, **12**, 347–78.

Garruto, R.M., Fukatsu, R., Yanagihara, R., Gajdusek, D.C., Hook, G. & Fiori, C.E. (1984). Imaging of calcium and aluminum in neurofibrillary tangle-bearing neurons in parkinsonism-dementia of Guam. *Proceedings of the National Academy of Sciences USA*, **81**, 1875–9.

Garruto, R.M., Gajdusek, D.C. & Chen, K-M. (1980). Amyotrophic lateral sclerosis among Chamorro migrants from Guam. *Annals of Neurology*, **8**, 612–19.

Garruto, R.M., Gajdusek, D.C. & Chen, K-M. (1981). Amyotrophic lateral sclerosis and parkinsonism-dementia among Filipino migrants to Guam. *Annals of Neurology*, **10**, 341–50.

Garruto, R.M., Shankar, S.K., Yanagihara, R., Salazar, A.M., Amyx, H.L. & Gajdusek, D.C. (1989). Low calcium, high aluminum diet-induced motor neuron pathology in cynomolgus monkeys. *Acta Neuropathologica*, **78**, 210–19.

Garruto, R.M., Yanagihara, R. & Gajdusek, D.C. (1985). Disappearance of high incidence amyotrophic lateral sclerosis and parkinsonism-dementia of Guam. *Neurology*, **35**, 193–8.

Garruto, R.M., Yanagihara, R. & Gajdusek, D.C. (1988). Cycads and amyotrophic lateral sclerosis/parkinsonism dementia. *Lancet*, **ii**, 1079.

Garruto, R.M. & Yase, Y. (1986). Neurodegenerative disorders of the western Pacific: the search for mechanisms of pathogenesis. *Trends in Neuroscience*, **9**, 368–74.

Gibbs, C.J., Jr. & Gajdusek, D.C. (1982). An update on long-term *in vivo* and *in vitro* studies designed to identify a virus as a cause of amyotrophic lateral sclerosis, parkinsonism-dementia and Parkinson disease. In *Advances in Neurology, Vol. 36, Human Motor Neuron Disease*, ed. L.P. Rowland, pp. 343–54. New York: Raven Press.

Golub, M.S., Gershwin, M.E., Donald, J.M., Negri, S. & Keen, C.L. (1987). Maternal and developmental toxicity of chronic aluminum exposure in mice. *Fundamental and Applied Toxicology*, **8**, 346–57.

Golub, M.S., Keen, C.L. & Gershwin, M.E. (1992). Neurodevelopmental effect of aluminum in mice: fostering studies. *Neurotoxicology and Teratology*, **14**, 177–82.

Gomez, M., Domingo, J.L. & Llobet, J.M. (1991). Developmental toxicity evaluation of oral aluminum in rats: influence of citrate. *Neurotoxicology and Teratology*, **13**, 323–8.

Gruskin, A.B. (1991). Aluminum toxicity in infants and children. In *Trace Elements in Nutrition of Children–II* (*Nestle Nutrition Workshop Series*, Vol. **23**), ed. R.K. Chandra, pp. 15–25. New York: Raven Press.

Guiroy, D.C., Mellini, M., Miyazaki, M., Hilbich, Safar, J., Garruto, R.M., Yanagihara, R., Beyreuther, K. & Gajdusek, D.C. (1993). Neurofibrillary tangles of Guamanian amyotrophic lateral sclerosis, parkinsonism-dementia and neurologically normal Guamanians contain a 4- to 4.5-kilodalton protein which is immunoreactive to anti-amyloid β/A4-protein antibodies. *Acta Neuropathologica*, **86**, 265–74.

Guiroy, D.C., Miyazaki, M., Multhaup, G., Fisher, P., Garruto, R.M., Beyreuther, K., Masters, C., Simms, G., Gibbs, C.J., Jr. & Gajdusek, D.C. (1987). Amyloid of neurofibrillary tangles of Guamanian parkinsonism-dementia and Alzheimer disease share an identical amino acid sequence. *Proceedings of the National Academy of Sciences USA*, **84**, 2073–7.

Harrington, C.R., Wischik, C.M., McArthur, F.K., Taylor, G.A., Edwardson, J.A. & Candy, J.M. (1994). Alzheimer's-disease like changes in tau protein processing: association with aluminum accumulation in brains of renal dialysis patients. *Lancet*, **343**, 993–7.

Hartwell, B.L. & Pember, F.R. (1918). The presence of aluminum as a reason for the difference in the effect of so-called acid soil on barley and rye. *Soil Science*, **6**, 259–79.

Hirano, A. (1973). Progress in the pathology of motor neuron diseases. In *Progress in Neuropathology*, vol. II, ed. H.M. Zimmerman, pp. 181–215. New York: Grune and Stratton.

Hirano, A., Kurland, L.T., Krooth, R.S. & Lessell, S. (1961a). Parkinsonism-dementia complex, an endemic disease on the island of Guam. I. Clinical features. *Brain*, **84**, 642–61.

Hirano, A., Malamud, N. & Kurland, L.T. (1961b). Parkinsonism-dementia complex, an endemic disease on the island of Guam. II. Pathological features. *Brain*, **84**, 662–79.

Hutchinson, G.E. (1943). The biogeochemistry of aluminum and of certain related elements. *Quarterly Review of Biology*, **18**, 1–29, 128–53, 242–62, 331–63.

Kidd, M. (1963). Paired helical filaments in electron microscopy in Alzheimer's disease. *Nature*, **197**, 192–3.

Kihara, T., Yoshida, S., Uebayashi, Y., Wakayama, I. & Yase, Y. (1994). Experimental model of motor neuron disease: oral aluminum neurotoxicity. *Biomedical Research*, **15**, 27–36.

Klein, G.L., Ott, S.M., Alfrey, A.C., Sherrard, D.J., Hazlet, T.K., Miller, N.L. & Maloney, N.A., Berquist, W.E., Ament, M.E. & Coburn, J.W. (1982a). Aluminum as a factor in the bone disease of long-term parenteral nutrition. *Transactions of the Association of American Physicians*, **95**, 155–64.

Klein, G.L., Alfrey, A.C., Miller, N.L., Sherrard, D.J., Hazlet, T.K., Ament, M.E. & Coburn, J.W. (1982b). Aluminum loading during total parenteral nutrition. *American Journal of Clinical Nutrition*, **35**, 1425–9.

Kurland, L.T. & Mulder, D.W. (1954). Epidemiologic investigations of amyotrophic lateral sclerosis. 1. Preliminary report on geographic distributions, with special reference to the Marianas Islands, including clinical and pathological observations. *Neurology*, **4**, 355–78, 438–48.

Kurland, L.T. & Mulder, D.W. (1955). Epidemiologic investigations of amyotrophic later sclerosis. 2. Familial aggregations indicative of dominant inheritance. *Neurology*, **5**, 182–96, 249–68.

Masters, C.L., Multhaup, G., Simms, G., Pottgiesser, J., Martins, R.N. & Beyreuther, K. (1985). Neuronal origin of a cerebral amyloid: neurofibrillary tangles of Alzheimer's disease contain the same protein as the amyloid of plaque cores and blood vessels. *EMBO Journal*, **4**, 2757–63.

Matsumoto, S., Hirano, A. & Goto, S. (1990a). Spinal cord neurofibrillary tangles of Guamanian amyotrophic lateral sclerosis and parkinsonism-dementia complex: An immunohistochemical study. *Neurology*, **40**, 975–9.

Matsumoto, S., Hirano, A. & Goto, S. (1990b). Ubiquitin-immunoreactive filamentous inclusions in anterior horn cells of Guamanian and non-Guamanian amyotrophic lateral sclerosis. *Acta Neuropathologica*, **80**, 233–8.

Misawa, T. & Shigeta, S. (1993). Effects of prenatal aluminum treatment on development and behavior in the rat. *Journal of Toxicological Sciences*, **18**, 43–8.

Molitoris, B.A., Froment, D.H., MacKenzie, T.A., Huffer, W.H. & Alfrey, A.C. (1989). Citrate: A major factor in the toxicity of orally administered aluminum compounds. *Kidney International*, **36**, 949–53.

Mulder, D.W., Kurland, L.T. & Irarte, L.L.G. (1954). Neurological diseases on the island of Guam. *United States Armed Forces Medical Journal*, **5**, 1724–39.

Muller, G., Bernuzzi, V., Desor, D., Hutin, M-F., Burnel, D. & Lehr, P.R. (1990). Developmental alterations in offspring of female rats orally intoxicated by aluminum lactate at different gestation periods. *Teratology*, **42**, 253–61.

Netter, P., Kessler, M., Gaucher, A. & Bannwarth, B. (1990). Does aluminium have a pathogenic role in dialysis associated arthropathy? *Annals of Rheumatic Disease*, **49**, 573–5.

Nikaido, T., Austin, J., Trueb, L. & Rinehart, R. (1972). Studies in ageing of the brain. II Microchemical analyses of the nervous system in Alzheimer patients. *Archives of Neurology*, **27**, 549–54.

Ott, S.M., Maloney, N.A., Klein, G.L., Alfrey, A.C., Ament, M.E., Coburn, J.W. & Sherrard, D.J. (1983). Aluminum is associated with low bone formation in patients receiving chronic parenteral nutrition. *Annals of Internal Medicine*, **98**, 910–14.

Owen, G.M. (1989). Aluminum-to-creatinine ratios in infancy. *Lancet*, **i**, 1397.

Parker, D.R., Zelazny, L.W. & Kinraide, T.B. (1989). Chemical speciation and plant toxicity to aqueous aluminum. In *Environmental Chemistry and Toxicology of Aluminium*, ed. T.E. Lewis, pp. 117–45. Chelsea, Michigan: Lewis Publishers.

Pavan, M.A., Bingham, F.T. & Pratt, P.F. (1982). Toxicity of aluminum to coffee in Ultisols and Oxisols amended with $CaCO_3$, $MgCO_3$ and $CaSO_4 \bullet 2H_2O$. *Soil Science Society of America Journal*, **46**, 1201–7.

Perl, D.P. & Brody, A.R. (1980). Alzheimer's disease: X-ray spectrometric evidence of aluminum accumulation in neurofibrillary tangle-bearing neurons. *Science*, **208**, 297–9.

Petit, T.L., Biederman, G.B., Jonas, P. & LeBoutillier, J.C. (1985). Neurobehavioral development following aluminum administration in infant rabbits. *Experimental Neurology*, **88**, 640–51.

Plato, C.C., Reed, D.M., Elizan, T.S. & Kurland, L.T. (1967). Amyotrophic lateral sclerosis/parkinsonism-dementia complex of Guam. IV. Familial and genetic investigations. *American Journal of Human Genetics*, **19**, 617–32.

Plato, C.C., Garruto, R.M., Fox, K.M. & Gajdusek, D.C. (1986). Amyotrophic lateral sclerosis and parkinsonism-dementia of Guam: a 25-year prospective case-control study. *American Journal of Epidemiology*, **124**, 643–56.

Rankin, J. & Manning, A. (1993). Alterations to the pattern of ultrasonic calling after prenatal exposure to aluminium sulfate. *Behavioral and Neural Biology*,

59, 136–42.

Roy, A.K., Sharma, A. & Talukder, G. (1988). Some aspects of aluminum toxicity in plants. *Botanical Review*, **54**, 145–78.

Schofield, C.L. & Trojnar, R.J. (1980). Aluminum toxicity to brook trout (Salvelinus fontinalis) in acidified waters. In *Polluted Rain*, ed. T.Y. Toribara, M.W. Miller and P.E. Morrow, pp. 341–66. New York: Plenum Press.

Scholtz, C.L., Swash, M., Gray, A., Kogeorgos, J. & Marsh, F. (1987). Neurofibrillary neuronal degeneration in dialysis dementia: a feature of aluminum toxicity. *Clinical Neuropathology*, **6**, 93–7.

Sedman, A. (1992). Aluminum toxicity in childhood. *Pediatric Nephrology*, **6**, 383–93.

Sedman, A.B., Klein, G.L., Merritt, R.J., Miller, N.L., Weber, K.O., Gill, W.L., Anand, H. & Alfrey, A.C. (1985). Evidence of aluminum loading in infants receiving intravenous therapy. *New England Journal of Medicine*, **312**, 1337–43.

Shankar, S., Yanagihara, R., Garruto, R.M., Grundke-Iqbal, I., Kosik, K.S. & Gajdusek, D.C. (1989). Immunocytochemical characterization of neurofibrillary tangles in amyotrophic lateral sclerosis and parkinsonism-dementia on Guam. *Annals of Neurology*, **25**, 146–51.

Shiraki, H. & Yase, Y. (1991). Amyotrophic lateral sclerosis and parkinsonism-dementia in the Kii Peninsula: comparison with the same disorders in Guam and with Alzheimer's disease. In *Handbook of Clinical Neurology-Diseases of the Motor System*, Vol. **59**, ed. J.M.B.V. De Jong, pp. 273–300. Amsterdam: Elsevier Scientific Publishing Company.

Shortle, W.C. & Smith, K.T. (1988). Aluminum-induced calcium deficiency syndrome in declining red spruce. *Science*, **240**, 1017–18.

Slanina, P., Frech, W., Ekström, L-G., Lööf, L., Slorach, S. & Cedergren, A. (1986). Dietary citric acid enhances absorption of aluminum in antacids. *Clinical Chemistry*, **32**, 539–41.

Strong, M.J. & Garruto, R.M. (1991). Neuron-specific thresholds of aluminum toxicity *in vitro*: a comparative analysis of dissociated fetal rabbit hippocampal and motor neuron-enriched cultures. *Laboratory Investigation*, **65**, 243–9.

Strong, M.J. & Garruto, R.M. (1994). Neuronal aging and age related disorders of the human nervous system. In *Biological Anthropology and Aging: Perspectives on Human Variation Over the Life Span*, ed. D.E. Crews and R.M. Garruto, pp. 214–31, New York: Oxford University Press.

Strong, M.J., Wolff, A.V., Wakayama, I. & Garruto, R.M. (1991). Aluminum-induced chronic myelopathy in rabbits. *Neurotoxicology*, **12**, 9–22.

Tillema, S. & Wijnberg, C.J. (1953). 'Endemic' amyotrophic lateral sclerosis on Guam; epidemiological data. A preliminary report. *Documente de Medi Geographica et Tropica*, **5**, 366–70.

Tsou, V.M., Young, R.M., Hart, M.H. & Vanderhoof, J.A. (1991). Elevated plasma aluminum levels in normal infants receiving antacids containing aluminum. *Pediatrics*, **87**, 148–51.

Wakayama, I., Kihara, T., Yoshida, S. & Garruto, R.M. (1993). Rare neuropil threads in amyotrophic lateral sclerosis and parkinsonism-dementia on Guam and in the Kii Peninsula of Japan. *Dementia*, **4**, 75–80.

Weberg, R. & Berstad, A. (1986). Gastrointestinal absorption of aluminium from single doses of aluminium containing antacids in man. *European Journal of*

Clinical Investigation, **16**, 428–32.

Wisniewski, H.M., Narang, H.K. & Terry, R.D. (1976). Neurofibrillary tangles of paired helical filaments. *Journal of the Neurological Sciences*, **27**, 173–81.

Wisniewski, H.M., Sturman, J. & Shenk, J.W. (1980). Aluminum chloride induced neurofibrillary changes in the developing rabbit: a chronic animal model. *Annals of Neurology*, **8**, 479–90.

Yano, I., Yoshida, S., Uebayashi, Y., Yoshimasu, F. & Yase, Y. (1989). Degenerative changes in the central nervous system of Japanese monkeys induced by oral administration of aluminum salt. *Biomedical Research*, **10**, 33–41.

Yase, Y. (1980). The role of aluminum in CNS degeneration with interaction of calcium. *Neurotoxicology*, **1**, 101–9.

Yasui, M., Yase, Y., Ota, K. & Garruto, R.M. (1991). Evaluation of magnesium, calcium, aluminum metabolism in rats and monkeys maintained on calcium-deficient diets. *Neurotoxicology*, **12**, 603–14.

Yokel, R.A. (1984). Toxicity of aluminum exposure during lactation to the maternal and suckling rabbit. *Toxicology and Applied Pharmacology*, **75**, 35–43.

Yokel, R.A. (1985). Toxicity of gestational aluminum exposure to the maternal rabbit and offspring. *Toxicology and Applied Pharmacology*, **79**, 121–33.

Yokel, R.A. (1987). Toxicity of aluminum exposure to the neonatal and immature rabbit. *Fundamental and Applied Toxicology*, **9**, 795–806.

Yokel, R.A. (1989). Aluminum produces age related behavioral toxicity in the rabbit. *Neurotoxicology and Tertology*, **11**, 237–42.

Index